Thomas Gauss

# Floquet Theory for a Class of Periodic Evolution Equations in an $L_p$-Setting

# Floquet Theory for a Class of Periodic Evolution Equations in an $L_p$-Setting

von
Thomas Gauss

KIT Scientific Publishing

Dissertation, Karlsruher Institut für Technologie
Fakultät für Mathematik, 2010

**Impressum**

Karlsruher Institut für Technologie (KIT)
KIT Scientific Publishing
Straße am Forum 2
D-76131 Karlsruhe
www.ksp.kit.edu

KIT – Universität des Landes Baden-Württemberg und nationales
Forschungszentrum in der Helmholtz-Gemeinschaft



KIT Scientific Publishing 2010
Print on Demand

ISBN 978-3-86644-542-0

# Floquet Theory for a Class of Periodic Evolution Equations in an $L_p$-Setting

Zur Erlangung des akademischen Grades eines

DOKTORS DER NATURWISSENSCHAFTEN

von der Fakultät für Mathematik der
Universität Karlsruhe (TH)
genehmigte

DISSERTATION

von

Dipl.-Math. Thomas Gauss,
geboren in Pforzheim

Tag der mündlichen Prüfung: 3. Februar 2010
Referent: Prof. Dr. Lutz Weis
Korreferent: Prof. Dr. Roland Schnaubelt

# Contents

# Introduction

In this thesis we explore the Floquet theory for a class of periodic evolution equations.

The motivating example arises from the following physical model of waveguides[1]. The (time-harmonic) electromagnetic field inside a *cylindrical waveguide* can be described by an equation of the form

$$J\frac{\partial}{\partial z}u(z) = M(z)u(z).$$

Here, we have assumed that the waveguide has a non-varying (bounded) cross section $\Omega \subset \mathbb{R}^2$ and its axis is the $z$-axis. Furthermore, the relevant material properties of the waveguide, namely the permittivity and the permeability, are assumed to be *periodic* w. r. t. the $z$-direction. The *Maxwell operator* $M$ then is a partial differential operator with periodic coefficients acting on the Hilbert space $H := L_2(\Omega) \times L_2(\Omega)$ and for a fixed $z$ the first and the second component of $u(z) \in H$ describes the electric and the magnetic field in each cross section $\Omega$, resp.. Finally, $J \in \mathscr{L}(H)$ is an invertible operator that can be represented as an (operator-valued) anti-diagonal matrix and thus "couples" both components of a solution $u$. We refer to [Der72], [Pru76] and the references therein for a more detailed explanation of the model.

Here, we will choose a more abstract setting to deal with the problem, namely we will consider nonautonomous evolution equations of the form

$$u'(t) + A_t u(t) = 0 \qquad (t \in \mathbb{R}) \tag{E}$$

where the operator family $(A_t)_{t\in\mathbb{R}}$ on a Banach space $X$ periodically depends on $t \in \mathbb{R}$ and the function $u$ takes values in $X$.

[1] The reader unfamiliar with the physical concept of waveguides (or electrodynamics in general) will find all necessary information in the monograph [Jac62].

Our central assumptions are that the operators $(A_t)_{t\in\mathbb{R}}$ have a common domain space $D$ which is compactly embedded into $X$ and that the resolvent sets all contain a common line parallel to the imaginary axis, e. g. $i\mathbb{R}$, and that the corresponding resolvents $(A_t-\lambda)^{-1}$ decay in the sense that $(|\lambda|+1)(A_t-\lambda)^{-1}$ is uniformly bounded for all $t\in\mathbb{R}$ (cf. condition ($A$-iv) on page 26).

As expected in the context of partial differential equations with periodic coefficients, *Bloch solutions*—and more generally *Floquet solutions*—will play an important role. As the central result we will obtain that all exponentially bounded solutions to (E) can be described as a superposition of a fixed family of Floquet solutions.

In particular, if $X$ is a (complex-valued) $L_q$-space we will prove that the so-called *Bloch property* holds for (E), i. e. the existence of a non-vanishing bounded solution implies the existence of a bounded Bloch solutions. We will also obtain the interesting result that the set of *Floquet exponents* coincides with the set of *Bloch exponents*, in other words if (E) has a Floquet solution then (E) also has a Bloch solution with the same exponent. In particular, this result could be used—e. g. when searching for so-called band gaps—to conclude the nonexistence of Floquet solutions already from the nonexistence of Bloch solutions.

In [Kuc93] Chapter 5 P. Kuchment gave such a description in a Hilbert space setting, i. e. the family $(A_t)_{t\in\mathbb{R}}$ acts on a Hilbert space $X$ and a solution $u$ is a locally square integrable $X$-valued function.

We extend P. Kuchment's result to a Banach space setting: We treat the case where $X$ is a *UMD-space* and allow the solutions to locally belong to $L_p$ for some fixed $p$ with $1<p<\infty$. Examples of UMD-spaces are all closed subspaces of $L_q$-spaces with $1<q<\infty$, in particular reflexive Sobolev and Hardy spaces.

Following P. Kuchment we use the Floquet transform to translate the problem into a corresponding statement about *analytic Fredholm homomorphisms of bundles*. The representation of exponentially bounded solutions as superpositions of Floquet solutions can then be obtained from abstract results of such homomorphisms due to M. Zaidenberg, S. Krein, P. Kuchment, A. Pankov and V. Palamodov.

The original treatise makes use of properties of Hilbert spaces that do not hold in general in the Banach space setting or even in $L_p$-spaces with $p \neq 2$. An important part of this thesis consists of replacing classical Hilbert space methods by modern methods of spectral theory and harmonic analysis.

In particular, the construction of regularizers needed to verify the Fredholm property mentioned above leads to a $R$-boundedness condition on the resolvents of $A_t$ that coincides with the boundedness condition of the original treatise if the new result is applied to a Hilbert space. The $R$-boundedness then allows to use results on operator-valued Fourier multipliers obtained by L. Weis, which provides the base for the extension to the Banach space setting. It is known that for many classes of differential operators $A_t$ this $R$-boundedness condition is satisfied, e. g. if the resolvent of $A_t$ or the semigroup generated by $A_t$ satisfies Gaussian bounds.

This thesis is structured as follows. In *Chapter 1* we will declare notations and state well-known or basic facts.

*Chapter 2* contains basic results on the operator family $(A_t)_{t \in \mathbb{T}}$ defined in the UMD-space setting that can be obtained more or less directly from the Hilbert space case.

The Fredholm property of the operator $\mathcal{L} = \partial + A_t$ on $L_p([0,1], X)$ will be proven in *Chapter 3* with the help of Fourier multiplier theorems.

In *Chapter 4* we study the dual situation and its relation with the operator $\mathcal{L}'$. Since Plancherel's theorem is not available if $X$ is not a Hilbert space we had to derive more general functional analytic duality arguments than the Hilbert space methods used in P. Kuchment's treatment.

*Chapter 5* generalizes the so-called *hypoellipticity* result—namely that quasiperiodic solutions have $L_p([0,1], D) \cap W_p^1([0,1], X)$ regularity—of P. Kuchment's treatise to the Banach space setting using the same methods as in Chapter 3 and Chapter 4. Furthermore, we extend the result from periodic functions to quasiperiodic functions that—thanks to the duality method used in Chapter 4—can now be treated analogously.

In *Chapter 6* we introduce and discuss the various notions of solutions relevant here.

In *Chapter 7* we reformulate our problem in terms of analytic Fredholm homomorphisms of bundles. We refer to Section 7.1 for a more detailed explanation of this transformation. In particular, we use a Banach space version of the *Floquet transform* as an isomorphism from the test function spaces to spaces of sections of bundles. Also, analogously as in the original treatise, the coincidence of Floquet and Bloch exponents holds.

Finally, the central result of this thesis, namely the representation of a solution as the superposition of Floquet solutions, will be given in *Chapter 8*.

For the convenience of the reader and for the sake of completeness, we provide a summary on the structures of bundles and sheaves that are used in Chapter 7 and Chapter 8 in an *appendix*.

### Acknowledgments

I am very grateful to my advisor, Professor Dr. Lutz Weis, for his guidance and support. Without his direction and supervision, this thesis would not have been possible.

I also thank Prof. Dr. Roland Schnaubelt for co-examining this thesis.

Finally, I'm thankful to my colleagues from the Research Training Group GRK 1294 and all members of the "Institut für Analysis" for creating a friendly and supportive atmosphere that always allowed an open discussion of all kinds of mathematical and non-mathematical problems.

The work on this thesis was supported by the German Research Foundation (DFG) through a grant in the research training group GRK 1294 "Analysis, Simulation and Design of Nanotechnological Processes".

# Chapter 1

# Notation and Definitions

## 1.1 Basic Notation

We denote by $\mathbb{N}$, $\mathbb{Z}$, $\mathbb{R}$ and $\mathbb{C}$ the natural[1], integer, real and complex numbers, resp.. Furthermore, we set $\mathbb{N}_0 := \mathbb{N} \cup \{0\}$.

For all $M \subset \mathbb{C}\backslash\{0\}$ we set $M^{-1} := \{ m^{-1} : m \in M \}$.

We will write $f := [M \ni m \mapsto \cdots \in N]$ to define (or denote) a function $f : M \to N$. We will sometimes omit the explicit specification of the domain $M$ or the codomain $N$ if it is clear out of the context.

To avoid confusion we remark that nowhere in this thesis $[\ldots]$ is used to denote an equivalence class.

For a function $f : M \to N$ and $m \in M$ we set $\delta_m f := f(m) \in N$.

If $f : M \to N$ is a function and $F$ is a family of functions defined on a set $m \subset M$ we will write $f \in F$ if $f_{|m} \in F$.

For a set $M$ we denote by $\mathbb{1}_M$ its characteristic function, i. e. $\mathbb{1}_M(m) = 1$ for all $m \in M$ and $\mathbb{1}_M(m) = 0$ if $m \notin M$.

$\lfloor\cdot\rfloor : \mathbb{R} \to \mathbb{Z}$ denotes the floor function, i. e. for all $t \in \mathbb{R}$ $\lfloor t \rfloor$ is defined to be the largest integer not greater than $t$.

For all $n, m \in \mathbb{Z}$ $\delta_{n,m}$ denotes the Kronecker delta, i. e. $\delta_{n,m} = 1$ if $n = m$ and $\delta_{n,m} = 0$ if $n \neq m$.

[1] i. e. the positive integers, in particular $0 \notin \mathbb{N}$

If $\Omega$ is a topological space we write $O \overset{\circ}{\subset} \Omega$ if $O \subset \Omega$ is open in $\Omega$. Analogously, we write $K \subset\subset \Omega$ if $K \subset \Omega$ is compact.

We denote by $B_\Omega(x, r)$ the open ball with radius $r$ around $x$ in a topological space $\Omega$.

A neighborhood of a point in a topological space is always assumed to be an open set.

Unless otherwise stated a subset of a topological space is endowed with the induced topology and the product of topological spaces is endowed with the product topology.

For all $n \in \mathbb{N}_0$ we will write

$\sum_{\alpha+\beta=n} \cdots$ shortly for $\sum_{(\alpha,\beta)\in\{\,(a,b)\in\mathbb{N}_0\times\mathbb{N}_0:\,a+b=n\,\}} \cdots$ and
$\sum_{\alpha+\beta+\gamma=n} \cdots$ shortly for $\sum_{(\alpha,\beta,\gamma)\in\{\,(a,b,c)\in\mathbb{N}_0\times\mathbb{N}_0\times\mathbb{N}_0:\,a+b+c=n\,\}} \cdots$.

We denote by $\mathbb{P}$ the space of polynomials in one variable with coefficients in $\mathbb{C}$.

## 1.2 Vector Spaces and Operators

We use the notion of *Fréchet space* as defined in [Trè67] Chapter 10, i. e. a Fréchet space is a metrizable, complete, locally convex, Hausdorff topological complex vector space. We will also use the equivalent description that a Fréchet space is a complex vector space endowed with a topology induced by a countable family of seminorms that is Hausdorff and complete (cf. [Sch80] § II.4).

We denote by $\mathscr{L}(X, Y)$ the space of continuous linear operators between the Fréchet spaces $X$ and $Y$ and as usual we set $\mathscr{L}(X) := \mathscr{L}(X, X)$.

We will use the well-known characterization of continuity of linear operators in terms of bounds of seminorms, cf. [Trè67] Proposition I.7.7.

If $X$ and $Y$ are Fréchet spaces and $A \in \mathscr{L}(X, Y)$ is bijective, then $A^{-1} \in \mathscr{L}(X, Y)$ and thus $A$ is an isomorphism of Fréchet spaces (see [Trè67] Corollary 1 to Theorem 17.1).

If, in particular, $X$ and $Y$ are Banach spaces[2], we equip $\mathscr{L}(X,Y)$ with the usual Banach space structure; in particular, its topology is induced by the (strong) operator norm.

Furthermore, we denote by $\mathscr{K}(X,Y)$ the space of linear compact operators between the Banach spaces $X$ and $Y$ and again, we set $\mathscr{K}(X) := \mathscr{K}(X,X)$.

$X \hookrightarrow Y$ means that the Banach space $X$ is continuously embedded in the Banach space $Y$ and $X \hookrightarrow\!\!\!\!\rightarrow Y$ analogously denotes a compact embedding.

We denote by $\rho(A)$ the resolvent set for any densely defined operator $A : X \supset D \to X$ on a Banach space $X$.

For Fréchet spaces $X_1$, $X_2$ and $Y$ we denote by $\mathscr{L}(X_1; X_2, Y)$ the space of continuous bilinear operators from $X_1 \times X_2$ to $Y$. By [Bou87] Corollary III.5.1 $A \in \mathscr{L}(X_1; X_2, Y)$ iff $[x_1 \mapsto A(x_1, x_2)] \in \mathscr{L}(X_1, Y)$ for all $x_2 \in X_2$ and $[x_2 \mapsto A(x_1, x_2)] \in \mathscr{L}(X_2, Y)$ for all $x_1 \in X_1$.

As usual, if $A \in \mathscr{L}(X)$ for some Fréchet space $X$ we set $A^0 := \mathrm{Id}_X$.

If $(a_k)_{k\in\mathbb{Z}} \subset X$ for some Banach space $X$ and both series $\sum_{k=0}^{\infty} a_k$ and $\sum_{k=1}^{\infty} a_{-k}$ converge absolutely we say $\sum_{k=-\infty}^{\infty} a_k$ converges absolutely and we set $\sum_{k=-\infty}^{\infty} a_k := \sum_{k=0}^{\infty} a_k + \sum_{k=1}^{\infty} a_{-k}$.

The following identity will be used passim and can be obtained by a simple calculation.

**1.2.1 Fact**
$\sum_{k=-\infty}^{\infty} \exp(-a|k|) = \frac{\exp(a)+1}{\exp(a)-1}$ for all $a > 0$.

## Duality

For a Fréchet space $X$, we denote by $X^* := \mathscr{L}(X, \mathbb{C})$ the dual space. We will call both the transpose of a continuous linear operator between Fréchet spaces (as defined in [Trè67] section 18 (2)) and the adjoint of a densely defined closed operator between Banach spaces (as defined in [Kat66] III.5.3) the *dual operator*. Clearly, the notion coincides in the case of bounded linear operators between Banach spaces. In any case, we will use the symbol $^*$ to denote the dual operator.

[2] which are always assumed to have a complex underlying vector space

If $X$ is a Fréchet space, $x \in X$ and $x' \in X^*$, we set $\langle x', x \rangle_X := x'(x)$.

If $X$ and $Y$ are Fréchet spaces and $A \in \mathscr{L}(X, Y)$ is bijective, then $A^* : X^* \to Y^*$ is an isomorphism of vector spaces, i. e. $A^*$ is linear and bijective (see [Trè67] Proposition 23.1).

If $X$ and $Y$ are Fréchet spaces and $A \in \mathscr{L}(X, Y)$ then we set $\operatorname{Coker} A := (\operatorname{Range} A)^\perp := \{ y' \in Y^* : y'(\operatorname{Range} A) = \{0\} \}$. Thus $\operatorname{Coker} A = \operatorname{Ker}(A^*)$.

**1.2.2 Fact**
If $X$ and $Y$ are Banach spaces and $A \in \mathscr{L}(X, Y)$ is a Fredholm operator, then $Y/_{(\operatorname{Range} A)} \cong \left(Y/_{(\operatorname{Range} A)}\right)^* \cong (\operatorname{Range} A)^\perp = \operatorname{Coker} A$, cf. [Wer05] Satz III.1.10.

Finally, we refer to e. g. [Kat66] as a general reference for basic properties of dual operators of closed operators between Banach spaces.

## 1.3 Standard Function Spaces

Throughout this section, let $a, b \in \mathbb{R}$ with $a < b$, $p \in (1, \infty)$ and $X$ be a Banach space.

We denote by $C[0,1]$ the Banach space of continuous complex-valued functions on $[0,1]$ endowed with the usual supremum norm $\|\cdot\|_\infty$. We denote by $C^1[0,1]$ the Banach space of 1-time continuously differentiable[3] complex-valued functions on $[0,1]$ endowed with the usual norm[4] $\|f\|_{C^1[0,1]} := \max(\|f\|_\infty, \|\partial f\|_\infty$.

More generally, $C(\mathbb{R}, X)$ denotes the space of continuous $X$-valued functions on the real line.

Furthermore, we denote by $C_c(\mathbb{R}, X)$ and $C_c^\infty(\mathbb{R}, X)$ the spaces of continuous and infinitely differentiable, resp., $X$-valued functions on the real line with compact support.

We denote by $L_p([a,b], X)$ and $L_p(\mathbb{R}, X)$ the corresponding Bochner space, i. e. the Banach space of (equivalence classes[5] of) $p$-integrable $X$-valued functions on $[a,b]$ and on $\mathbb{R}$, resp., endowed with the usual norm $\|\cdot\|_p$, and by

[3] Of course, in the points 0 and 1 one-sided differentiability is meant.

[4] $\partial f$ denotes the derivative of $f$.

[5] We identify two functions if they coincide almost everywhere w. r. t. the Lebesgue measure. We follow the usual convention of abuse of notation and—whenever there is no confusion about the

$W_p^1([a,b],X)$ the corresponding Sobolev space, i.e. the spaces of (equivalence classes of) 1-time weakly differentiable $p$-integrable $X$-valued functions on $(a,b)$ endowed with the usual norm[6]

$$\|f\|_{W_p^1([a,b],X)} := \left(\|f\|^p_{L_p([a,b],X)} + \|\partial f\|^p_{L_p([a,b],X)}\right)^{1/p}$$

as defined in [Ama95] Section III.1.1.

Let $t \in [a,b]$. For each $\langle f\rangle \in W_p^1([a,b],X)$ there is a unique continuous representant $f : [a,b] \to X$ and we set $\delta_t\langle f\rangle := \langle f\rangle(t) := f(t)$. Then $\delta_t \in \mathscr{L}((W_p^1([a,b],X)),X)$, cf. [Ama95] Section III.1.4.

We denote by $L_{p,\mathrm{loc}}(\mathbb{R},X)$ and $W^1_{p,\mathrm{loc}}(\mathbb{R},X)$ the local Bochner and Sobolev spaces, i.e. (an equivalence class of) a function $f : \mathbb{R} \to X$ belongs to $L_{p,\mathrm{loc}}(\mathbb{R},X)$ or $W^1_{p,\mathrm{loc}}(\mathbb{R},X)$, resp., iff for all $\alpha,\beta \in \mathbb{R}$ with $\alpha < \beta$ the restriction $f_{|[\alpha,\beta]}$ belongs to $L_p([\alpha,\beta],X)$ or $W_p^1([\alpha,\beta],X)$, resp.. We remark that [HP57] Theorem 3.5.4 (3) implies that those functions are strongly measurable.

We note that for all $t \in (a,b)$ $f \in W_p^1([a,b],X)$ iff $f \in W_p^1([a,t],X)$, $f \in W_p^1([t,b],X)$ and for the continuous representants $f_{|[a,t]}$ and $f_{|[t,b]}(t)$ $f_{|[a,t]}(t) = f_{|[t,b]}(t)$ holds.

## 1.4 Quasiperiodic Functions

We remark that throughout this thesis, the (quasi-)period of a (quasi-)periodic function is always 1.

During this section, let $z \in \mathbb{C}\backslash\{0\}$, $p \in (1,\infty)$ and $X$ be a Banach space. As in the previous section, equivalence of functions is again understood as coincidence almost everywhere.

We say, a function $g : \mathbb{R} \to X$ is $z$-quasiperiodic, if $g(\xi+1) = zg(\xi)$ for all $\xi \in \mathbb{R}$. It is obvious that for every function $f : [0,1] \to X$ with $f(1) = zf(0)$ there is a unique $z$-quasiperiodic extension, which we will denote by $\mathrm{E}_z f$. Conversely, every $z$-quasiperiodic function is of the form $\mathrm{E}_z f$, where $f : [0,1] \to X$ is a function with $f(1) = zf(0)$.

meaning—do not distinguish in notation between an equivalence class of a function and a (fixed) representant.

[6]Throughout this thesis, the derivative of a (at least) weakly differentiable function $f$ will be denoted by $\partial f$.

We say, an equivalence class $\langle g\rangle$ (of functions on $\mathbb{R}$ to $X$) is $z$-quasiperiodic, if for a (or equivalently for all) representant $g : \mathbb{R} \to X$ $g(\xi+1) = zg(\xi)$ for almost all $\xi \in \mathbb{R}$. If $N \subset \mathbb{R}$ denotes the corresponding null set for a fixed representant $g$, i. e. $g(\xi+1) = zg(\xi)$ for all $\xi \in \mathbb{R}\setminus N$, then $\tilde{N} := \bigcup_{k\in\mathbb{Z}}(k+N) \supset N$ is a null set such that $t \in \mathbb{R}\setminus\tilde{N}$ implies $t+k \in \mathbb{R}\setminus\tilde{N}$ for all $k \in \mathbb{Z}$. We say, $\tilde{N}$ is a *quasiperiodicity null set* for the representant $g$.

Again, it is obvious that for every equivalence class $\langle f\rangle$ of functions from $[0,1]$ to $X$ there is a unique extension to a $z$-quasiperiodic equivalence class of functions on $\mathbb{R}$ to $X$, which, by abuse of notation, we will denote by $\mathrm{E}_z\langle f\rangle$. Conversely, every $z$-quasiperiodic equivalence class of functions on $\mathbb{R}$ to $X$ is of the form $\mathrm{E}_z\langle f\rangle$, where $\langle f\rangle$ is an equivalence class of functions from $[0,1]$ to $X$. $\mathrm{E}_z\langle f\rangle$ has a continuous representant iff $\langle f\rangle$ has a continuous representant $f : [0,1] \to X$ with $f(1) = zf(0)$.

We set $W_p^1([0,1],X)_z := \{\, f \in W_p^1([0,1],X) : f(1) = zf(0)\,\}$. Since
$W_p^1([0,1],X)_z$
is the kernel of $z\delta_0 - \delta_1$ on $W_p^1([0,1],X)$ the space $W_p^1([0,1],X)_z$ is a closed subspace of $W_p^1([0,1],X)$ and therefore a Banach space.

Finally, we denote by $C^\infty([0,1],X)_z$ the space of restrictions to $[0,1]$ of infinitely differentiable $z$-quasiperiodic $X$-valued functions on $\mathbb{R}$.

## Periodic Functions

In particular, we will deal with periodic, i. e. 1-quasiperiodic, functions.

We will also write $\mathrm{E}_\mathbb{T}$ instead of $\mathrm{E}_1$ and $C^\infty(\mathbb{T},X)$ instead of $C^\infty([0,1],X)_1$. Furthermore, we denote by $C(\mathbb{T},X)$ the space of restrictions to $[0,1]$ of continuous periodic $X$-valued functions on $\mathbb{R}$.

We set $L_p(\mathbb{T},X) := \mathrm{E}_\mathbb{T}(L_p([0,1],X))$. We call the reader's attention to the fact that (equivalence classes of) functions in $L_p(\mathbb{T},X)$ are defined $\mathbb{R}$ in contrast to, e. g.[7], $C(\mathbb{T},X)$ that only contains functions defined on $[0,1]$. Clearly, $L_p(\mathbb{T},X) \subset L_{p,\mathrm{loc}}(\mathbb{R},X)$.

Finally, we remark that $f \in W_p^1([0,1],X)_1$ iff there exists $g \in L_p([0,1],X)$

[7]Actually, all other function spaces in this thesis that are associated with the symbol $\mathbb{T}$ contain only (equivalence classes of) functions defined on $[0,1]$.

such that $\int_0^1 g\phi\,dt = -\int_0^1 f\partial\phi\,dt$ for all $\phi \in C^\infty(\mathbb{T},\mathbb{C})$. In this case we have $\partial f = g$ in the weak sense.

Another characterization of $W_p^1([0,1],X)_1$ using Fourier coefficients can be found in [AB02] Lemma 2.1.

## 1.5 Analyticity

Let $\Omega \overset{\circ}{\subset} \mathbb{C}$ and $X$ be a Fréchet space or the dual of a Fréchet space endowed with the weak-* topology, i. e. the topology of pointwise convergence on the predual.

We say, $f : \Omega \to X$ is *analytic* if it is differentiable at each point of $\Omega$, i. e. for each $z \in \Omega$ the limit $\lim\limits_{\xi\to z} \frac{f(\xi)-f(z)}{\xi - z}$ exists.

We denote by $A(\Omega, X)$ the space of analytic functions on $\Omega$ with values in $X$.

We remark that $X$ is quasi-complete (cf. [Sch80] Section IV.6.1 for the case that $X$ is the dual of a Fréchet space) and thus the closed, convex, circled hull of any compact subset of $X$ is compact (cf. [Sch80] II.4.3 Corollary). In that case, analyticity implies continuity and coincides with weak analyticity, cf. [Gro53] Théorème § 2.1:[8]

**1.5.1 Fact**
$A(\Omega, X) \subset C(\Omega, X)$.

**1.5.2 Fact** (*Weak Analyticity*)
$f \in A(\Omega, X)$ iff $x' \circ f \in A(\Omega, \mathbb{C})$ for all $x' \in X^*$.

**1.5.3 Fact**
Let $X$ be a Fréchet space and endow $X^*$ with the weak-* topology.
Then $f \in A(\Omega, X^*)$ iff $[z \mapsto \langle f(z), x\rangle_X] \in A(\Omega, \mathbb{C})$ for all $x \in X$.

We also recall the results from [Gro53] Remarque § 2.4.

**1.5.4 Fact**
$A(\Omega, X)$ together with pointwise addition and scalar multiplication is a $\mathbb{C}$-vector space.

---

[8] In the case that $X$ is the dual of a Fréchet space, say $X = Y^*$, $X^*$ here denotes the topological dual of $X$ endowed with the strong topology. $X^*$ then coincides with $Y$, cf. [FW68] Satz § 14.1.3.

**1.5.5 Fact** (*Product Rule*)
Let $X$, $Y$ and $Z$ be Fréchet spaces and $\pi \in \mathscr{L}(X;Y,Z)$.
Then $[\Omega \ni z \mapsto \pi(f(z),g(z))] \in A(\Omega,Z)$ for all $f \in A(\Omega,X)$ and $g \in A(\Omega,Y)$

For the rest of this section we assume that $X$ is a Banach space.

**1.5.6 Definition and Fact** (*Fréchet Space Structure on $A(\Omega,X)$*)
We endow $A(\Omega,X)$ with the compact-open topology, i. e. the topology generated by all seminorms of the form $A(\Omega,X) \ni f \mapsto \sup_{x\in K} \|f(x)\|_X$ where $\emptyset \neq K \subset\subset \Omega$ (Cf. [Cha85] Section 16.8 and [Dug70] Theorem XII.7.2.). Then $A(\Omega,X)$ is a Fréchet space, cf. [Cha85] Theorem 16.13.

Analogously to the scalar case, the Weierstraß convergence theorem (cf. [RS02] Theorem 8.4.1) holds. We state the following consequence.

**1.5.7 Fact**
$\partial \in \mathscr{L}(A(\Omega,X))$.

As direct consequence of the compact-open topology and Fact 1.5.3 we obtain:

**1.5.8 Fact**
$\delta_z \in (A(\Omega,\mathbb{C}))^*$ for each $z \in \Omega$.
If we endow $(A(\Omega,\mathbb{C}))^*$ with the weak-* topology then
$$[z \mapsto \delta_z] \in A(\Omega,(A(\Omega,\mathbb{C}))^*).$$

It can be easily shown that the product defined in Fact 1.5.5 is continuous in the following sense.

**1.5.9 Fact**
Let $Y$ and $Z$ be Banach spaces and $\pi \in \mathscr{L}(X;Y,Z)$.
For all $f \in A(\Omega,X)$ and $g \in A(\Omega,Y)$ we set $\Pi(f,g) := [\Omega \ni z \mapsto \pi(f(z),g(z))]$. Then $\Pi \in \mathscr{L}(A(\Omega,X);A(\Omega,Y),A(\Omega,E))$.

We also state the following two applications of Fact 1.5.9.

**1.5.10 Fact**
Let $Y$ and $Z$ be Banach spaces, $A \in A(\Omega,\mathscr{L}(X,Y))$, $B \in A(\Omega,\mathscr{L}(Y,Z))$ and $f \in A(\Omega,X)$. Then $[x \mapsto \big(A(x)\big)f(x)] \in A(\Omega,Y)$ and $[x \mapsto \big(B(x)\big)\big(A(x)\big)] \in A(\Omega,\mathscr{L}(X,Z)$.

**1.5.11 Fact**
Let $\alpha \in A(\Omega, C[0,1])$ and $f \in A(\Omega, L_p([0,1], X))$.
Then $\alpha f \in A(\Omega, L_p([0,1], X))$.

As a direct consequence of Fact 1.5.2 and the Hahn-Banach theorem we obtain:

**1.5.12 Fact**
Let $f \in A(\Omega, X)$ with $f(\Omega) \subset U$, where $U$ is a closed subspace of $X$.
Then $f \in A(\Omega, U)$.

## Series Expansion

We will make use of the following well-known expansions, cf. [DS58] Section III.14.

**1.5.13 Definition and Fact** (*Power series*)
Let $f \in A(\Omega, X)$.
For each $z \in \Omega$ and all $r > 0$ such that $\overline{B_{\mathbb{C}}(z,r)} \subset \Omega$, $f$ can be *expanded* into a *power series (about $z$)*. I. e. there exists $(x_k)_{k \in \mathbb{N}_0} \subset X$ such that the series $\sum_{k=0}^{\infty} x_k(\cdot - z)^k$ converges absolutely and uniformly to $f$ on $\overline{B_{\mathbb{C}}(z,r)}$. The coefficients $(x_k)_{k \in \mathbb{N}_0} \subset X$ are uniquely determined and are given by $x_k = (\partial^k f)(z)/k!$ for each $k \in \mathbb{N}_0$.

If, in particular, $\Omega = B_{\mathbb{C}}(z,r)$ for some $r > 0$ and $z \in \mathbb{C}$ then there exists a uniquely determined power series $\sum_{k=0}^{\infty} x_k(\cdot - z)^k$ about $z$ that converges absolutely and pointwise to $f$ on $\Omega$ and $\left[\Omega \ni \xi \mapsto \sum_{k=0}^{N} x_k(\xi - z)^k\right] \overset{N\to\infty}{\longrightarrow} f$ in $A(\Omega, X)$.

Conversely, if $g : \Omega \to X$ can be locally expanded into power series, i. e. if for any $z \in \Omega$ there exists $r > 0$ with $B_{\mathbb{C}}(z,r) \subset \Omega$ and a power series $\sum_{k=0}^{\infty} x_k(\cdot - z)^k$ that converges absolutely pointwise to $g$ on $B$, then $g \in A(\Omega, X)$ and the power series expansion about $z$ of $g$ coincides with
$\sum_{k=0}^{\infty} x_k(\cdot - z)^k$.

**1.5.14 Remark**
The expansion into a unique power series also holds for each $f \in A(\Omega, Y)$ where $Y$ is a Fréchet space, cf. [Gro53] Théorème § 2.1. In particular, let $F \in A(\Omega, A(\Omega, X))$ and $n \in \mathbb{N}_0$. If we denote by $\sum_{k=0}^{\infty} f_k^{(z)}(\cdot - z)^k$ the power series expansion of $F(z) \in A(\Omega, X)$ about $z$, in other words $f_k^{(z)} :=$

$\delta_z(\partial^k(F(z)))/k! \in X$ for each $k \in \mathbb{N}_0$, then Fact 1.5.8 in combination with Fact 1.5.7 yields $[z \mapsto f_n^{(z)}] \in A(\Omega, X)$.

**1.5.15 Definition and Fact** (*Laurent series, Cauchy-Hadamard formula*)
Let $\Omega_{\alpha,\beta} := \{ z \in \mathbb{C} : \alpha < |z| < \beta \}$ (where $\alpha \geq 0$ and $\beta \in (\alpha, \infty]$ be an annulus with center 0.
Every $f \in A(\Omega_{\alpha,\beta}, X)$ has a unique *Laurent series expansion (with center* 0*)*, i. e. there exists (a uniquely determined) $(x_k)_{k\in\mathbb{Z}} \subset X$ such that the series $\sum_{k=-\infty}^{\infty} x_k(\cdot)^k$ converges absolutely and pointwise to $f$ on $\Omega_{\alpha,\beta}$. In particular, $\limsup_{k\to\infty} \left(\|x_{-k}\|_X\right)^{1/k} \leq \alpha$ and[9] $\beta \leq \left(\limsup_{k\to\infty} \left(\|x_k\|_X\right)^{1/k}\right)^{-1} \in \mathbb{R} \cup \{\infty\}$. Furthermore, the convergence above is uniform on every $K \subset\subset \mathbb{C}\backslash\{0\}$, in other words $\left[\Omega_{\alpha,\beta} \ni \xi \mapsto \sum_{k=-N}^{N} x_k(\xi)^k\right] \underset{A(\Omega_{\alpha,\beta},X)}{\overset{N\to\infty}{\longrightarrow}} f$.

Conversely, if $(x_k)_{k\in\mathbb{Z}} \subset X$ such that

$$\alpha = \limsup_{k\to\infty} \left(\|x_{-k}\|_X\right)^{1/k} < \beta = \left(\limsup_{k\to\infty} \left(\|x_k\|_X\right)^{1/k}\right)^{-1} \in \mathbb{R} \cup \{\infty\}$$

then $\sum_{k=-\infty}^{\infty} x_k(\cdot)^k$ converges absolutely pointwise on $\Omega_{\alpha,\beta}$, to say $g$. Furthermore, $g \in A(\Omega_{\alpha,\beta}, X)$ and $\sum_{k=-\infty}^{\infty} x_k(\cdot)^k$ is the Laurent series expansion of $g$.

## Analytic Functions on Banach Spaces

Analytic functions defined on Banach spaces will only occur in compositions. The following basic statements will be all we need and we refer to the monograph [Cha85] for a detailed treatise of the topic. In particular, we refer to [Cha85] Theorem 14.13 for an equivalent definition and [Cha85] Theorem 5.9 for a proof of Fact 1.5.16.

Let $U \overset{\circ}{\subset} X$ and $Y$ be a Banach space. We say, $f : U \to Y$ is *analytic* if it is Fréchet differentiable on $U$. (We remind the reader that all Banach spaces in this thesis are complex vector spaces.)

We denote by $A(U, Y)$ the set of all analytic functions from $U$ to $Y$.

[9] Here, we use the convention $1/\infty := 0$, of course.

**1.5.16 Fact**
Let $Y$ and $Z$ be Banach spaces, $U \overset{\circ}{\subset} X$ and $V \overset{\circ}{\subset} Y$.
If $f \in A(U,Y)$ with $f(U) \subset V$ and $g \in A(V,Z)$ then $g \circ f \in A(U,Z)$.

A direct calculation finally yields the following result.

**1.5.17 Fact**
Let $X \in \{C[0,1], C^1[0,1]\}$.
We define $\exp : X \to X$ by $\exp f := \big[[0,1] \ni t \mapsto \exp(f(t))\big]$ for each $f \in X$.
Then a simple calculation yields $\exp \in A(X,X)$ and for each $f \in X$ the derivative $\partial \exp f \in \mathscr{L}(X)$ is given by $X \ni g \mapsto (\exp f) \cdot g$.

### Analytic Sets

**1.5.18 Definition and Fact**
A subset $Z$ of $\Omega$ is called an *analytic set in $\Omega$* if for each $z \in \Omega$ there exist a neighborhood $O \overset{\circ}{\subset} \Omega$ of $z$, $l \in \mathbb{N}$ and analytic functions $f_i \in A(\Omega, \mathbb{C})$ for each $i = 1, \ldots, l$ such that $Z \cap O = \{\, \xi \in \Omega : \; f_1(\xi) = \ldots f_l(\xi) = 0 \,\}$.
Every analytic set is closed in $\Omega$, cf. e. g. [Łoj91] Section II.§ 3.4.
Clearly, an analytic set in $\mathbb{C}\backslash\{0\}$ is either $\mathbb{C}\backslash\{0\}$ or a discrete set of points.

## 1.6 Intervals and Distance on the Torus

The following technical definition is used in connection with the domain of periodic functions.

An open $\mathbb{T}$-interval of length $\delta < 1$ is a subset of $[0,1]$ of the form $[0,1] \cap (I + \mathbb{Z})$, where $I \subset \mathbb{R}$ is an open interval of length $\delta$. The center of such a $\mathbb{T}$-interval is (the unique point) $[0,1] \cap (c + \mathbb{Z})$, where $c$ is the center of $I$.

For all $a, b \in \mathbb{R}$ we define their $\mathbb{T}$-distance $\mathrm{d}_{\mathbb{T}}(a,b) := \min_{n \in \mathbb{Z}} |a + n - b|$.

## 1.7 Fourier Series of Banach Space Valued Functions

Again, let $X$ be a Banach space. We denote by

$$s(\mathbb{Z}, X) := \{\, (x_k)_{k \in \mathbb{Z}} \subset X : \; \lim_{|k| \to \infty} \|k^n x_k\|_X = 0 \text{ for all } n \in \mathbb{N} \,\}$$

the space of rapidly decreasing sequences with values in $X$ (which takes the role of the Schwartz space).

We define $\mathcal{F}_X : C^\infty(\mathbb{T}, X) \to s(\mathbb{Z}, X)$ by

$$\mathcal{F}_X f := ((\mathcal{F}_X f)_k)_{k\in\mathbb{Z}} := (\int_0^1 f(t)e^{-2\pi ikt}\,dt)_{k\in\mathbb{Z}}.$$

Then $\mathcal{F}_X$ is well-defined, bijective and the inverse $\mathcal{F}_X^{-1} : s(\mathbb{Z}, X) \to C^\infty(\mathbb{T}, X)$ is given by $\mathcal{F}_X^{-1}(c_k)_{k\in\mathbb{Z}} = [t \mapsto \sum_{k=-\infty}^{\infty} c_k e^{2\pi ikt}]$.

## 1.8 Multiplication Operators

Throughout this section, let $p \in (1, \infty)$ and $X$ be a Banach space.

**1.8.1 Definition**
Let $M$ be a set, $\alpha : M \to \mathbb{C}$ and $X$ a Banach space. For any function $f : M \to X$ we set $\mathrm{M}[\alpha]f := [M \ni m \mapsto \alpha(m)f(m)]$. If it is clear from the context that $\alpha$ acts as such a multiplication operator we also shortly write $\alpha f$ instead of $\mathrm{M}[\alpha]f$.

We directly obtain the following statements.

**1.8.2 Fact**
$[\alpha \mapsto \mathrm{M}[\alpha]] \in \mathscr{L}(C[0,1], \mathscr{L}(L_p([0,1], X))$.
In particular, $\mathrm{M}[\alpha] \in \mathscr{L}(L_p([0,1], X))$ for each $\alpha \in C[0,1]$.

**1.8.3 Fact**
$[\alpha \mapsto \mathrm{M}[\alpha]] \in \mathscr{L}(C^1[0,1], \mathscr{L}(W_p^1([0,1], X))$.
In particular, $\mathrm{M}[\alpha] \in \mathscr{L}(W_p^1([0,1], X))$ for each $\alpha \in C^1[0,1]$.

## 1.9 Complex Power Functions

**Motivation**
We will need analytic logarithm and power[10] functions (locally) for all points in $z \in \mathbb{C}\backslash\{0\}$. For each point $z \in \mathbb{C}\backslash\{0\}$ we choose an arbitrary (but from then on fixed) branch of a logarithm that is defined on a suitable neighborhood of $z$. (For the sake of clarity we also will provide an index to indicate to which point

[10] for real-valued exponents

a logarithm function "belongs", e. g. $\log_{(1/2)} \frac{3}{4}$ will denote the evaluation at $\frac{3}{4}$ of the branch associated with $1/2$ and e. g. $\log_{(1)} \frac{3}{4}$ will denote the evaluation at $\frac{3}{4}$ of the branch associated with 1 (the values may not coincide).) For each point $z \in \mathbb{C}\backslash\{0\}$ we will then use this branch of the logarithm to define a corresponding power function, again on a suitable neighborhood of $z$. △

We will now provide the technical details and some simple facts. Additionally, we refer to [FL94] §§ V.1 and V.2..

**Construction**
Let $z \in \mathbb{C}\backslash\{0\}$.

We set $\mathbb{B}_z := B_{\mathbb{C}}(z, |z|)$.

Then there is an analytic logarithm function $\xi \mapsto \log_{(z)} \xi$ on $\mathbb{B}_z$, i. e. $\log_{(z)} \in A(\mathbb{B}_z, \mathbb{C})$ and $\exp(\log_{(z)} \xi)) = \xi$ for all $\xi \in \mathbb{B}_z$.

For all $\xi \in \mathbb{B}_z$ and $t \in \mathbb{R}$ we set $\xi^t_{(z)} := \exp(t \log_{(z)} \xi)$. We write shortly $\xi^{(\cdot)}_{(z)}$ and $\xi^{(-\cdot)}_{(z)}$ instead of $\mathbb{R} \ni t \mapsto \xi^t_{(z)}$ and $\mathbb{R} \ni t \mapsto \xi^{-t}_{(z)}$, resp.. △

The following statements hold for all $z \in \mathbb{C}\backslash\{0\}$.

**1.9.1 Fact**
$[\xi \mapsto \xi^{(\cdot)}_{(z)}], [\xi \mapsto \xi^{(-\cdot)}_{(z)}] \in A(\mathbb{B}_z, C[0,1])$.

$[\xi \mapsto \xi^{(\cdot)}_{(z)}], [\xi \mapsto \xi^{(-\cdot)}_{(z)}] \in A(\mathbb{B}_z, C^1[0,1])$.

(This is a simple consequence of Fact 1.5.2, Fact 1.5.17 and Fact 1.5.16.)

**1.9.2 Fact**
For every $\xi \in \mathbb{B}_z$ and $n \in \mathbb{Z}$ $\xi^n_{(z)}$ do not depend on $z$, namely: If $n > 0$ $\xi^n_{(z)} = \prod_{j=1}^n \xi$, if $n < 0$ $\xi^n_{(z)} = 1/\xi^n_{(z)}$ and $\xi^0_{(z)} = 1$. Therefore we will sometimes omit the subscript in case of an integer-valued exponent.

**1.9.3 Fact**
$\xi^{s+t}_{(z)} = \xi^s_{(z)} \xi^t_{(z)}$ for every $\xi \in \mathbb{B}_z$ and $s, t \in \mathbb{R}$. In particular $\xi^{(\cdot)}_{(z)} \xi^{(-\cdot)}_{(z)} = \xi^{(-\cdot)}_{(z)} \xi^{(\cdot)}_{(z)} = \mathbb{1}_{\mathbb{B}_z}$ for every $\xi \in \mathbb{B}_z$.

## 1.10 UMD-spaces, $R$-boundedness

A Banach space $X$ is called a *UMD-space* if the Hilbert transform, defined on the Schwartz space $\mathcal{S}(\mathbb{R}, X)$, can be extended to a bounded operator on

$L_p(\mathbb{R}, X)$ for some (or, equivalently, for all) $p \in (1, \infty)$. We refer to, e.g., [Ama95] § III.4.4 for the details and basic properties. In particular we will use, that if $X$ is a UMD-space and $Y$ is a Banach space that is isomorphic to $X$, then $Y$ is a UMD-space as well. Furthermore, every UMD-space is reflexive.

We remark that every Hilbert space is a UMD-space. Furthermore, if $\Omega, \mu$ is a $\sigma$-finite measure space then every closed subspace and every quotient space[11] of $L_p(\Omega, \mu)$ for each $p \in (1, \infty)$ is a UMD-space, cf. [Ama95] Theorem III.4.5.2. In particular, all reflexive Sobolev and Hardy space are UMD-spaces.

If $X$ and $Y$ are Banach spaces a family of operators $\mathcal{A} \subset \mathscr{L}(X, Y)$ is called *R-bounded* if there exists $c > 0$ such that for all $n \in \mathbb{N}$, $A_1, \ldots, A_n \in \mathcal{A}$ and $x_1, \ldots, x_n \in X$

$$\| \textstyle\sum_{i=1}^n r_i A_i x_i \|_{L_p([0,1],Y)} \leq c \| \textstyle\sum_{i=1}^n r_i x_i \|_{L_p([0,1],X)}$$

for some (or, equivalently, for all) $p \in (1, \infty)$. Here, $r_n := \big[[0,1] \ni t \mapsto \operatorname{sign} \sin(2^n \pi t)\big]$ for all $n \in \mathbb{N}$ denote the Rademacher functions. (If we want to emphasize that $\mathcal{A} \subset \mathscr{L}(X, Y)$ we write $(X, Y)$-*R-bounded.*)

In that case we will denote by $R_{(X,Y)}(\mathcal{A})$ the smallest $c$ such that the above inequality holds. It will be clear from the context which $p$ is meant and we therefore omit a corresponding indication of the dependence on $p$.

If $H$ is a Hilbert space, all bounded sets in $\mathscr{L}(H)$ are $R$-bounded.

If $X$ is an $L_q(\Omega)$-space with $1 \leq q < \infty$ one can show that $(X, X)$-$R$-boundedness is equivalent to the following square function estimate

$$\| (\textstyle\sum_{i=1}^n |A_i x_i|^2)^{1/2} \|_{L_q(\Omega)} \leq c \| (\textstyle\sum_{i=1}^n |x_i|^2)^{1/2} \|_{L_q(\Omega)}.$$

More details on both definitions, basic properties and a remark on the meaning of the "$R$" can be found in [KW04] Section I.2.

[11] by a closed subspace

# Chapter 2

# Basic Framework

The following notations will be used throughout this thesis.

Let $1 < p < \infty$, $q := (1 - 1/p)^{-1}$ and $X$ be a UMD-space.

## 2.1 The Operator Family $(A_t)_{t\in\mathbb{T}}$

For all $t \in \mathbb{R}$ let $A_t : X \supset D(A_t) \to X$ be a closed operator such that

($A$-i) there exists a normed space $(D, \|\cdot\|_D)$ such that $D(A_t) = D$ for all $t \in \mathbb{R}$ and (the set) $D$ is a dense subspace of $X$ and

($A$-ii) $\|\cdot\|_D$ is, uniformly in $t$, equivalent to all graph norms $\|\cdot\|_{A_t}$ (where $\|d\|_{A_t} := \|d\|_X + \|A_t d\|_X$ for all $d \in D$), i. e. there exists $c_D > 0$ such that $c_D^{-1}\|d\|_D \le \|d\|_{A_t} \le c_D\|d\|_D$ for all $d \in D$ and all $t \in \mathbb{R}$.

Since $A_t$ (for, say, $t := 0$) is closed, $D$ is a Banach space. Clearly, $D \hookrightarrow X$ and $A_t \in \mathscr{L}(D, X)$ for all $t \in \mathbb{R}$.

Furthermore, we assume that $\mathbb{R} \ni t \mapsto A_t$ is periodic, i. e. $A_t = A_{t+1}$ for all $t \in \mathbb{R}$. To remind the reader of periodicity, we often write $t \in \mathbb{T}$ instead of $t \in \mathbb{R}$ when referring to indices of the operator family $(A_t)_{t\in\mathbb{R}}$.

## 2.2 The Lifted Operator $\mathcal{A}$

For a $D$-valued function $f$ defined on $M \subset \mathbb{R}$ we set

$$\mathcal{A}f := [M \ni t \mapsto A_t(f(t))].$$

We also assume (throughout this thesis) that

($A$-iii) $[t \mapsto A_t] \in C(\mathbb{T}, \mathscr{L}(D,X))$

holds.[1]

Obviously, this implies $\mathcal{A}f \in L_p([0,1],X)$ for all $f \in L_p([0,1],D)$.

The "realization" $\mathcal{A} : L_p([0,1],X) \supset L_p([0,1],D) \to L_p([0,1],X)$ then is a closed operator since its graph norm on its domain is equivalent to the norm of (the Banach space) $L_p([0,1],D)$. Furthermore, $\mathcal{A}$ is densely defined.

## 2.3 Constant Families $(A_{t_0})_{t\in\mathbb{T}}$, $\mathcal{A}_{t_0}$

We remark that for (a fixed) $t_0 \in \mathbb{T}$ the (constant) family

$$(A_{t_0} : X \supset D \to X)_{t\in\mathbb{T}}$$

fulfills the conditions ($A$-i), ($A$-ii) and ($A$-iii). Analogously to Section 2.2, we denote by $\mathcal{A}_{t_0}f$ the corresponding map $t \mapsto A_{t_0}(f(t))$ for a $D$-valued function $f$. Thus, again we obtain a closed, densely defined operator $\mathcal{A}_{t_0} : L_p([0,1],X) \supset L_p([0,1],D) \to L_p([0,1],X)$.

## 2.4 $\mathcal{W}_1[a,b]$, $\mathcal{W}_{1,z}[0,1]$, $\mathcal{W}_0[a,b]$, $\mathcal{W}_{-1,z}[0,1]$, $\mathcal{W}_1(\mathbb{T})$

During this section, let $z \in \mathbb{C}\backslash\{0\}$ and $a, b \in \mathbb{R}$ with $a < b$.

We will now introduce the Banach spaces $\mathcal{W}_1[a,b]$, $\mathcal{W}_{1,z}[0,1]$, $\mathcal{W}_0[a,b]$ and $\mathcal{W}_{-1,z}[0,1]$ where the first index should remind the reader of (weak) differentiability and the second one of quasiperiodicity.

For the definition of the intersection and sum of normed spaces and their corresponding natural norms we refer to [Tri95] Section 1.2.1.

We endow

$$\mathcal{W}_1[a,b] := L_p([a,b],D) \cap W_p^1([a,b],X)$$

with the norm $\|f\|_{\mathcal{W}_1[a,b]} := \left(\|f\|^p_{L_p([a,b],D)} + \|\partial f\|^p_{L_p([a,b],X)}\right)^{1/p}$. Then $\mathcal{W}_1[a,b]$ is a Banach space (and its norm is equivalent to the natural intersection norm $\|f\|_\cap := \max\{\|f\|_{L_p([a,b],D)}, \|f\|_{W_p^1([a,b],X)}\}$).

[1] Clearly, this is equivalent to $[t \mapsto A_t] \in C(\mathbb{R}, \mathscr{L}(D,X))$.

Similarly, we define the subspace of $z$-quasiperiodic functions[2]

$$\mathcal{W}_{1,z}[0,1] := L_p([0,1],D) \cap W_p^1([0,1],X)_z$$

of $\mathcal{W}_1[0,1]$ and we set $\|f\|_{\mathcal{W}_{1,z}[0,1]} := \|f\|_{\mathcal{W}_1[0,1]}$ for all $f \in \mathcal{W}_{1,z}[0,1]$. Then $\mathcal{W}_{1,z}[0,1]$ is a Banach space (and its norm is equivalent to the natural intersection norm $\|f\|_\cap := \max\{\|f\|_{L_p([0,1],D)}, \|f\|_{W_p^1([0,1],X)_z}\}$).

We set

$$\mathcal{W}_0[a,b] := L_p([a,b],X).$$

Furthermore, we define

$$\mathcal{W}_{-1,z}[0,1] := \big(\mathcal{W}_{1,1/z}[0,1]\big)^*.$$

We call the reader's attention to the index "$1/z$" which, of course, only serves as a more intuitive presentation due to duality. By [Yos71] Proposition 1.6 we get the representation $\mathcal{W}_{-1,z}[0,1] = \big(L_p([0,1],D) \cap W_p^1([0,1],X)_{1/z}\big)^* \cong \big(L_p([0,1],D)\big)^* + \big(W_p^1([0,1],X)_{1/z}\big)^*$.

For intuitive reasons, we set

$$\mathcal{W}_1(\mathbb{T}) := \mathcal{W}_{1,1}[0,1].$$

As a direct consequence of the continuous point evaluation in $W_p^1([a,b],X))$ (cf. Section 1.3), we obtain:

**2.4.1 Fact**

$\delta_t \in \mathscr{L}(\mathcal{W}_1[0,1],X)$ for all $t \in [0,1]$.

Finally, a mollifying argument (cf. [Ama95] Section III.4.2) easily yields:

**2.4.2 Fact**

$C^\infty([0,1],D)_z$ is dense in $\mathcal{W}_{1,z}[0,1]$.

[2] Cf. Section 1.4.

# Chapter 3

# Fredholm Property of $\mathcal{L}$

We set $\mathcal{L}_{\mathcal{W}_{1,z}[0,1]} := \partial + \mathcal{A} \in \mathscr{L}(\mathcal{W}_{1,z}[0,1], \mathcal{W}_0[0,1])$ for all $z \in \mathbb{C}\backslash\{0\}$ (and, in particular, $\mathcal{L}_{\mathcal{W}_1(\mathbb{T})} := \partial + \mathcal{A} \in \mathscr{L}(\mathcal{W}_1(\mathbb{T}), \mathcal{W}_0[0,1])$).

Before we impose further restrictions on (the dual operators of) $(A_t)_{t\in\mathbb{T}}$ we will now show that under suitable conditions $\mathcal{L}_{\mathcal{W}_{1,z}[0,1]}$ is a Fredholm operator.

The following theorem (or its corollary, resp.) provides the basis for constructing regularizers[1].

**3.1.3 Theorem**
Assume that for (a fixed) $t_0 \in \mathbb{T}$ there exists $\rho \in \mathbb{R}$ such that $\rho + i\mathbb{R} \subset \rho(A_{t_0})$ and $\{\, (|\lambda|+1)(A_{t_0} - \lambda)^{-1} : \lambda \in \rho + i\mathbb{R} \,\}$ is $(X,X)$-$R$-bounded. Furthermore, we define $\mathcal{L}_{t_0} \in \mathscr{L}(\mathcal{W}_1(\mathbb{T}), \mathcal{W}_0[0,1])$ by $\mathcal{L}_{t_0} := \partial + \mathcal{A}_{t_0} - \rho$.
Then $\mathcal{L}_{t_0}$ is invertible, i. e. $\mathcal{L}_{t_0}$ has a bounded inverse $\mathcal{B}_{t_0} \in \mathscr{L}(\mathcal{W}_0[0,1], \mathcal{W}_1(\mathbb{T}))$.

*Remarks on the proof.*
The statement mainly follows from [AB02] Theorem 2.3 and is an application of the Marcinkiewicz multiplier theorem. However, for the convenience of the reader we represent the complete proof in the current situation.

We also remark that the assumptions directly yield that $D$ is also a UMD-space, since $(A_{t_0} - \rho)^{-1} : X \to D$ is an isomorphism.

Finally, we remind the reader that by $\mathcal{F}_X : C^\infty(\mathbb{T}, X) \to s(\mathbb{Z}, X)$ and analogously $\mathcal{F}_D$ we refer to the Fourier transform as defined in Section 1.7.

*Proof.*
For all $f \in C^\infty(\mathbb{T}, X)$ let $\mathcal{B}_{t_0} f := \mathcal{F}_D^{-1} M \mathcal{F}_X f$ where $M : s(\mathbb{Z}, X) \to s(\mathbb{Z}, D)$

[1] Cf. the remarks on the proof of Theorem 3.1.6.

is defined by $M(c_k)_{k\in\mathbb{Z}} := (M_k c_k)_{k\in\mathbb{Z}} := ((A_{t_0} - \lambda_k)^{-1} c_k)_{k\in\mathbb{Z}}$ with $\lambda_k := \rho - 2\pi k i \in \rho(A_{t_0})$ for each $k \in \mathbb{Z}$. We remark that $M$ is well-defined because of the $R$-boundedness condition on the resolvents in combination with the identity $A_{t_0}(A_{t_0} - \lambda)^{-1} = \lambda(A_{t_0} - \lambda)^{-1} + \mathrm{Id}$ for all $\lambda \in \rho(A_{t_0})$.

We will prove in a moment that $\mathcal{B}_{t_0}$ extends to bounded linear operator from $\mathcal{W}_0[0,1]$ to $\mathcal{W}_1(\mathbb{T})$ which we will denote by the same symbol.

Before, we show first how invertibility of $\mathcal{L}_{t_0}$ follows from this: A direct calculation shows that $\mathcal{L}_{t_0}\mathcal{F}_D^{-1}(c_k)_{k\in\mathbb{Z}} = \mathcal{F}_X^{-1}((A_{t_0} - \lambda_k)c_k)_{k\in\mathbb{Z}}$ for all $(c_k)_{k\in\mathbb{Z}} \in s(\mathbb{Z}, D)$. Thus for all $f \in C^\infty(\mathbb{T}, X)$

$$\mathcal{L}_{t_0}\mathcal{B}_{t_0}f = \mathcal{L}_{t_0}\mathcal{F}_D^{-1}M\mathcal{F}_X f = \mathcal{F}_X^{-1}((A_{t_0} - \lambda_k)(A_{t_0} - \lambda_k)^{-1}(\mathcal{F}_X f)_k)_{k\in\mathbb{Z}} = \mathcal{F}_X^{-1}((\mathcal{F}_X f)_k)_{k\in\mathbb{Z}} = f.$$

On the other hand,

$$\mathcal{F}_X\mathcal{L}_{t_0}f = ((A_{t_0} - \lambda_k)(\mathcal{F}_D f)_k)_{k\in\mathbb{Z}} \text{ for all } f \in C^\infty(\mathbb{T}, D)$$

and therefore

$$\mathcal{B}_{t_0}\mathcal{L}_{t_0}f = \mathcal{F}_D^{-1}M\mathcal{F}_X\mathcal{L}_{t_0}f = \mathcal{F}_D^{-1}M((A_{t_0} - \lambda_k)(\mathcal{F}_D f)_k)_{k\in\mathbb{Z}} = \mathcal{F}_D^{-1}((A_{t_0} - \lambda_k)^{-1}(A_{t_0} - \lambda_k)(\mathcal{F}_D f)_k)_{k\in\mathbb{Z}} = \mathcal{F}_D^{-1}((\mathcal{F}_D f)_k)_{k\in\mathbb{Z}} = f.$$

Since $\mathcal{L}_{t_0} \in \mathscr{L}(\mathcal{W}_1(\mathbb{T}), \mathcal{W}_0[0,1])$ and $\mathcal{B}_{t_0} \in \mathscr{L}(\mathcal{W}_0[0,1], \mathcal{W}_1(\mathbb{T}))$ by density[2] of $C^\infty(\mathbb{T}, D)$ in $\mathcal{W}_1(\mathbb{T})$ and density of $C^\infty(\mathbb{T}, X)$ in $\mathcal{W}_0[0,1]$, resp., it follows that $\mathcal{L}_{t_0}$ is invertible.

It remains to show that indeed $\mathcal{B}_{t_0} \in \mathscr{L}(\mathcal{W}_0[0,1], \mathcal{W}_1(\mathbb{T}))$. First we note, that $\mathcal{B}_{t_0}(C^\infty(\mathbb{T}, X)) \subset C^\infty(\mathbb{T}, D) \subset \mathcal{W}_1(\mathbb{T})$. Thus by density of $C^\infty(\mathbb{T}, X)$ in $\mathcal{W}_0[0,1]$ it suffices to show that[3] $\|\mathcal{B}_{t_0}f\|_{\mathcal{W}_1(\mathbb{T})} \lesssim \|f\|_{\mathcal{W}_0[0,1]}$ for all $f \in C^\infty(\mathbb{T}, X)$. To this end we will show that (a) $\|\mathcal{B}_{t_0}f\|_{L_p(\mathbb{T},D)} \lesssim \|f\|_{\mathcal{W}_0[0,1]}$ and (b) $\|\partial\mathcal{B}_{t_0}f\|_{L_p(\mathbb{T},X)} \lesssim \|f\|_{\mathcal{W}_0[0,1]}$ for all $f \in C^\infty(\mathbb{T}, X)$.

For all $(c_k)_{k\in\mathbb{Z}} \in s(\mathbb{Z}, D)$ obviously

$$\partial\mathcal{F}_D^{-1}(c_k)_{k\in\mathbb{Z}} = \mathcal{F}_D^{-1}\tilde{M}(c_k)_{k\in\mathbb{Z}} = \mathcal{F}_X^{-1}\tilde{M}(c_k)_{k\in\mathbb{Z}}$$

where $\tilde{M}(c_k)_{k\in\mathbb{Z}} := (\tilde{M}_k c_k)_{k\in\mathbb{Z}} := (2\pi i k c_k)_{k\in\mathbb{Z}}$. Thus[4] $\partial\mathcal{B}_{t_0}f = \mathcal{F}_X^{-1}\tilde{M}M\mathcal{F}_X f$ for all $f \in C^\infty(\mathbb{T}, X)$.

Thus (a) and (b) are equivalent to

---

[2] Cf. Fact 2.4.2.

[3] We use the symbol $\lesssim$ in the sense that there exists some $C > 0$ such that $\|\mathcal{B}_{t_0}f\|_{\mathcal{W}_1(\mathbb{T})} \leq C\|f\|_{\mathcal{W}_0[0,1]}$ for all $f \in C^\infty(\mathbb{T}, X)$. The further occurrences of said symbol during this proof are meant analogously.

[4] Of course, $\partial\mathcal{F}_D^{-1}(c_k)_{k\in\mathbb{Z}} = \mathcal{F}_D^{-1}\tilde{M}(c_k)_{k\in\mathbb{Z}}$ and thus $\partial\mathcal{B}_{t_0}f = \mathcal{F}_D^{-1}\tilde{M}M\mathcal{F}_X f$ also holds. However, in general $\tilde{M}M$ fails to be a Fourier multiplier between $D$ and $X$.

$\|\mathcal{F}_D^{-1} M (c_k)_{k\in\mathbb{Z}}\|_{L_p(\mathbb{T},D)} \lesssim \|\mathcal{F}_X^{-1}(c_k)_{k\in\mathbb{Z}}\|_{\mathcal{W}_0[0,1]}$ and
$\|\mathcal{F}_X^{-1}\tilde{M}M(c_k)_{k\in\mathbb{Z}}\|_{L_p(\mathbb{T},X)} \lesssim \|\mathcal{F}_X^{-1}(c_k)_{k\in\mathbb{Z}}\|_{\mathcal{W}_0[0,1]}$
for all $(c_k)_{k\in\mathbb{Z}} \in \mathcal{F}_X(C^\infty(\mathbb{T},X)) = s(\mathbb{Z},X)$, i.e. $M$ and $\tilde{M}M$ are Fourier multipliers. By [AB02] Theorem 1.3 it suffices to show that

(a') the sets $\{\, k[M_{k+1} - M_k] : k \in \mathbb{Z}\,\}$ and $\{\, M_k : k \in \mathbb{Z}\,\}$ are $(X,D)$-$R$-bounded and that

(b') the sets

$$\{\, k[\tilde{M}_{k+1}M_{k+1} - \tilde{M}_k M_k] : k \in \mathbb{Z}\,\} = \{\, k[2\pi i(k+1)M_{k+1} - 2\pi i k M_k] : k \in \mathbb{Z}\,\}$$

and

$$\{\, \tilde{M}_k M_k : k \in \mathbb{Z}\,\} = \{\, 2\pi i k M_k : k \in \mathbb{Z}\,\}$$

are $(X,X)$-$R$-bounded.

By Kahane's contraction principle[5] we obtain

$$R_{(X,X)}(\{\, 2\pi i k M_k : k \in \mathbb{Z}\,\}) \leq 2R$$

with $R := R_{(X,X)}(\{\, (1+|\lambda|)(A_{t_0} - \lambda)^{-1} : \lambda \in \rho + i\mathbb{R}\,\}) < \infty$. We already remark for a later use that, similarly, we get

$$R_{(X,X)}(\{\, 2\pi i k M_{k+1} : k \in \mathbb{Z}\,\}) \leq (2+2\pi)R$$

and thus

$$R_{(X,X)}(\{\, k M_k : k \in \mathbb{Z}\,\}) \leq 2R/(2\pi) = \pi^{-1}R$$

and

$$R_{(X,X)}(\{\, k M_{k+1} : k \in \mathbb{Z}\,\}) \leq (2+2\pi)R/(2\pi) = (\pi^{-1}+1)R.$$

The resolvent identity yields

$$k[2\pi i(k+1)M_{k+1} - 2\pi i k M_k] = 2\pi i k M_{k+1} - 2\pi i k M_k 2\pi i k M_{k+1}.$$

Hence by [KW04] Fact 2.8

$$\begin{aligned}
&R_{(X,X)}(\{\, k[2\pi i(k+1)M_{k+1} - 2\pi i k M_k] : k \in \mathbb{Z}\,\}) \leq \\
&R_{(X,X)}(\{\, 2\pi i k M_{k+1} : k \in \mathbb{Z}\,\}) + \\
&\quad R_{(X,X)}(\{\, 2\pi i k M_k : k \in \mathbb{Z}\,\}) R_{(X,X)}(\{\, 2\pi i k M_{k+1} : k \in \mathbb{Z}\,\}) \leq \\
&(2+2\pi)R + 2R(2+2\pi)R.
\end{aligned}$$

Therefore (b') holds.

Again, by the identity $A_{t_0}(A_{t_0} - \lambda)^{-1} = \lambda(A_{t_0} - \lambda)^{-1} + \mathrm{Id}$ for all $\lambda \in \rho(A_{t_0})$) we get

[5] See, e. g., [KW04] Proposition 2.5.

$$R_{(X,D)}(\{ M_k : k \in \mathbb{Z} \}) \leq$$
$$c_D\big(R_{(X,X)}(\{ M_k : k \in \mathbb{Z} \}) + R_{(X,X)}(\{ \lambda_k M_k : k \in \mathbb{Z} \}) + 1\big)$$

where $c_D$ is the constant given by condition ($A$-ii). Another application of Kahane's contraction principle now yields $R_{(X,X)}(\{ M_k : k \in \mathbb{Z} \}) \leq R$ and $R_{(X,X)}(\{ \lambda_k M_k : k \in \mathbb{Z} \}) \leq 2R$. Therefore $\{ M_k : k \in \mathbb{Z} \}$ is $(X, D)$-$R$-bounded.

Finally, with the same argument we obtain

$$R_{(X,D)}(\{ k[M_{k+1} - M_k] : k \in \mathbb{Z} \}) \leq$$
$$c_D\big(R_{(X,X)}(\{ k[M_{k+1} - M_k] : k \in \mathbb{Z} \}) +$$
$$R_{(X,X)}(\{ k[\lambda_{k+1} M_{k+1} - \lambda_k M_k] : k \in \mathbb{Z} \})\big).$$

Furthermore, [KW04] Fact 2.8 yields

$$R_{(X,X)}(\{ k[M_{k+1} - M_k] : k \in \mathbb{Z} \}) \leq$$
$$R_{(X,X)}(\{ kM_{k+1} : k \in \mathbb{Z} \}) + R_{(X,X)}(\{ kM_k : k \in \mathbb{Z} \}) \leq$$
$$(\pi^{-1} + 1)R + \pi^{-1}R = (2\pi^{-1} + 1)R$$

and using again the resolvent identity we get

$$R_{(X,X)}(\{ k[\lambda_{k+1} M_{k+1} - \lambda_k M_k] : k \in \mathbb{Z} \}) \leq$$
$$R_{(X,X)}(\{ 2\pi ik M_{k+1} : k \in \mathbb{Z} \}) +$$
$$R_{(X,X)}(\{ \lambda_k M_k : k \in \mathbb{Z} \}) R_{(X,X)}(\{ 2\pi ik M_{k+1} : k \in \mathbb{Z} \}) \leq$$
$$(2 + 2\pi)R + 2R(2 + 2\pi)R.$$

Therefore (a') holds. This finishes the proof. □

### 3.1.4 Corollary

Assume that for the family $(A_t)_{t\in\mathbb{T}}$ the following condition holds.

($A$-iv) There exists $\rho \in \mathbb{R}$ such that $\rho + i\mathbb{R} \subset \rho(A_t)$ and
$$\{ (|\lambda| + 1)(A_t - \lambda)^{-1} : \lambda \in \rho + i\mathbb{R} \}$$
is uniformly $(X, X)$-$R$-bounded for all $t \in \mathbb{T}$, i. e. there exists $c_R > 0$ such that $R_{(X,X)}(\{ (|\lambda| + 1)(A_{t_0} - \lambda)^{-1} : \lambda \in \rho + i\mathbb{R} \}) < c_R$ for all $t \in \mathbb{T}$.

Then $\{ \mathcal{B}_t : t \in \mathbb{T} \}$, where $\mathcal{B}_t$ is the inverse of $\mathcal{L}_t$ according to Theorem 3.1.3, is bounded (in $\mathscr{L}(\mathcal{W}_0[0,1], \mathcal{W}_1(\mathbb{T}))$).

*Proof.*
We recall that in the proof of Theorem 3.1.3 we have shown boundedness of $\mathcal{B}_{t_0}$ by estimating certain $R$-bounds from above (by[6] $c_D((2\pi^{-1} + 2\pi + 3)R +$

[6]Here, of course, we are using the notations from the proof of Theorem 3.1.3. We remark that

$(4+4\pi)R^2)$, $c_D(3R+1)$, $(2+2\pi)R+(4+4\pi)R^2$ and $2R$, resp.) in order to apply [AB02] Theorem 1.3. Using condition ($A$-iv) those bounds obviously can be estimated from above independently of $t$. An examination of the proof of the cited theorem now yields that the norm bound of $\mathcal{B}_t$ then also can be estimated from above independently of $t$. □

As a final ingredient to establish the Fredholm property of $\mathcal{L}_{\mathcal{W}_1(\mathbb{T})}$ we will need a compact embedding of $\mathcal{W}_1(\mathbb{T})$ into $\mathcal{W}_0[0,1]$. This can be obtained by introducing an assumption on the domains $D$, namely we will use the following result by Aubin (cf. [Aub63]).

**3.1.5 Fact**
Assume that

($A$-v) $D \hookrightarrow\!\!\!\!\rightarrow X$

holds.
Then $\mathcal{W}_1(\mathbb{T}) \hookrightarrow\!\!\!\!\rightarrow \mathcal{W}_0[0,1]$.

**3.1.6 Theorem**
Let ($A$-iv) and ($A$-v) hold.
Then $\mathcal{L}_{\mathcal{W}_1(\mathbb{T})}$ is a Fredholm operator.

*Remarks on the proof.*
We imitate the proof of [Kuc93] Theorem 5.1.4 and will construct a left- and a right-regularizer, i. e. operators $\mathcal{R}_{\mathrm{L}}, \mathcal{R}_{\mathrm{R}} \in \mathscr{L}(\mathcal{W}_0[0,1], \mathcal{W}_1(\mathbb{T}))$ such that $\mathcal{R}_{\mathrm{L}}\mathcal{L}_{\mathcal{W}_1(\mathbb{T})} - \mathrm{Id}_{\mathcal{W}_1(\mathbb{T})}$ and $\mathcal{L}_{\mathcal{W}_1(\mathbb{T})}\mathcal{R}_{\mathrm{R}} - \mathrm{Id}_{\mathcal{W}_0[0,1]}$ are compact operators. Then [Sch73] Theorem 2.1. yields that $\mathcal{L}_{\mathcal{W}_1(\mathbb{T})}$ is a Fredholm operator which will prove the theorem. We remark that we will here use the notions to the torus w. r. t. $\mathbb{T}$ that we introduced in Section 1.6.

*Proof.*
For a (at first fixed) $1/2 > \delta > 0$ let $(U_j)_{j=1,\dots,N}$ be an open cover of $[0,1]$ by $N = N(\delta)$ open $\mathbb{T}$-intervals of the length $2\delta$ such that any point of $[0,1]$ is covered at most twice and for each $j = 1,\dots,N$ we denote by $t_j$ the center of $U_j$. Let[7] $(\phi_j)_{j=1,\dots,N} \subset C^\infty(\mathbb{T},[0,1])$ be a partition of unity subordinate[8] to this cover. Furthermore, for each $j = 1,\dots,N$ let $\psi_j \in C^\infty(\mathbb{T},[0,1])$ such that $\operatorname{supp}\psi_j \subset U_j$ and $\psi_j \equiv 1$ on $\operatorname{supp}\phi_j$ and let $\widetilde{\psi}_j \in C^\infty(\mathbb{T},[0,1])$ such that

in general $R$ depends on $t_0$.

[7] Of course, $C^\infty(\mathbb{T},[0,1])$ means the subset of all $[0,1]$-valued functions of $C^\infty(\mathbb{T},\mathbb{C})$.

[8] I. e. $\operatorname{supp}\phi_j \subset U_j$ for each $j = 1,\dots,N$.

$\operatorname{supp}\widetilde{\psi}_j \subset U_j$ and $\widetilde{\psi}_j \equiv 1$ on $\operatorname{supp}\psi_j$ for all $j = 1, \ldots, N$.

By Theorem 3.1.3 $\mathcal{L}_t := \partial + \mathcal{A}_t - \rho \in \mathscr{L}(\mathcal{W}_1(\mathbb{T}), \mathcal{W}_0[0,1])$ has a bounded inverse $\mathcal{B}_t \in \mathscr{L}(\mathcal{W}_0[0,1], \mathcal{W}_1(\mathbb{T}))$ for each $t \in [0,1]$ and Corollary 3.1.4 yields that the norms of the operators $\{\mathcal{B}_t\}_{t\in[0,1]}$ can be estimated by a common constant $c_{\mathcal{B}} > 0$ independently of $t$.

We set $\mathcal{B}f := \sum_{j=1}^N \mathrm{M}[\phi_j]\mathcal{B}_{t_j}\mathrm{M}[\psi_j]f$ for each $f \in \mathcal{W}_0[0,1]$. By Fact 1.8.2 and Fact 1.8.3 we obtain $\mathrm{M}[\phi_j]\mathcal{B}_{t_j}\mathrm{M}[\psi_j] \in \mathscr{L}(\mathcal{W}_0[0,1], \mathcal{W}_1(\mathbb{T}))$ for each $j = 1, \ldots, N$ and thus by Fact 3.1.5 $\mathrm{M}[\phi_j]\mathcal{B}_{t_j}\mathrm{M}[\psi_j] \in \mathscr{K}(\mathcal{W}_0[0,1])$. Therefore

$$\mathcal{B} \in \mathscr{L}(\mathcal{W}_0[0,1], \mathcal{W}_1(\mathbb{T})) \text{ and } \mathcal{B} \in \mathscr{K}(\mathcal{W}_0[0,1]).$$

A simple calculation shows

$$\mathcal{L}_{\mathcal{W}_1(\mathbb{T})}\mathcal{B} = \sum_{j=1}^N \mathrm{M}[\phi_j]\mathcal{L}_{t_j}\mathcal{B}_{t_j}\mathrm{M}[\psi_j] + \sum_{j=1}^N \mathrm{M}[\partial\phi_j]\mathcal{B}_{t_j}\mathrm{M}[\psi_j] + \rho\sum_{j=1}^N \mathrm{M}[\phi_j]\mathcal{B}_{t_j}\mathrm{M}[\psi_j] + \sum_{j=1}^N \mathrm{M}[\phi_j]\big(\mathcal{A} - \mathcal{A}_{t_j}\big)\mathcal{B}_{t_j}\mathrm{M}[\psi_j] \text{ on } \mathcal{W}_0[0,1].$$

For the first summand the definition directly yields

$$\sum_{j=1}^N \mathrm{M}[\phi_j]\mathcal{L}_{t_j}\mathcal{B}_{t_j}\mathrm{M}[\psi_j] = \mathrm{Id}_{\mathcal{W}_0[0,1]}.$$

As above, for the second summand

$$\mathcal{K}_1 := \sum_{j=1}^N \mathrm{M}[\partial\phi_j]\mathcal{B}_{t_j}\mathrm{M}[\psi_j] \in \mathscr{K}(\mathcal{W}_0[0,1])$$

and clearly for the third summand

$$\mathcal{K}_2 := \rho\sum_{j=1}^N \mathrm{M}[\phi_j]\mathcal{B}_{t_j}\mathrm{M}[\psi_j] = \rho\mathcal{B} \in \mathscr{K}(\mathcal{W}_0[0,1])$$

holds.

We will now show that the last summand $\mathcal{S} := \sum_{j=1}^N \phi_j\big(\mathcal{A} - \mathcal{A}_{t_j}\big)\mathcal{B}_{t_j}\psi_j$ is a contraction in $\mathcal{W}_0[0,1]$.

We denote by $\omega(\delta) := \sup\{\, \|A_t - A_s\|_{\mathscr{L}(D,X)} : t, s \in [0,1], \mathrm{d}_{\mathbb{T}}(t,s) \leq \delta \,\}$ the modulus of continuity of $\mathbb{T} \ni t \mapsto A_t \in \mathscr{L}(D,X)$. In combination with periodicity of the family $(A_t)_{t\in\mathbb{T}}$ we obtain by a compactness argument $\omega(\delta) \xrightarrow{\delta\to 0} 0$. Note that $\mathbb{1}_{U_j}(t)\|A_t - A_{t_j}\|_{\mathscr{L}(D,X)} \leq \omega(\delta)$ for all $j = 1, \ldots, N$ and all $t \in [0,1]$ since all points of $U_j$ have at most $\mathbb{T}$-distance $\delta$ to the center of $U_j$.

Furthermore, by the inequalities of Jensen and Hölder

$a^p + b^p \leq (a+b)^p \leq 2^{p-1}(a^p + b^p)$ for all $a, b \geq 0$.

Thus, if for each $j = 1, \ldots, N$ $g_j : [0,1] \to X$ is a function with $\operatorname{supp} g_j \in U_j$ then

$$\sum_{j=1}^{N} \|g_j(t)\|_X^p \leq \big(\sum_{j=1}^{N} \|g_j(t)\|_X\big)^p \leq 2^{p-1} \sum_{j=1}^{N} \|g_j(t)\|_X^p \text{ for each } t \in [0,1],$$

since by construction of the sets $U_j$ at a fixed point all but at most two summands vanish.

Therefore we obtain for all $f \in \mathcal{W}_0[0,1]$

$$\|\mathcal{S}f\|^p_{\mathcal{W}_0[0,1]} = \Big\| \sum_{j=1}^{N} \mathrm{M}[\phi_j]\big(\mathcal{A} - \mathcal{A}_{t_j}\big)\mathcal{B}_{t_j}\mathrm{M}[\psi_j]f \Big\|^p_{\mathcal{W}_0[0,1]} =$$

$$\int_0^1 \Big\| \sum_{j=1}^{N} \big(\mathrm{M}[\phi_j]\big(\mathcal{A} - \mathcal{A}_{t_j}\big)\mathcal{B}_{t_j}\mathrm{M}[\psi_j]f\big)(t) \Big\|^p_X \, dt \leq$$

$$\int_0^1 \Big( \sum_{j=1}^{N} \big\| \big(\mathrm{M}[\phi_j]\big(\mathcal{A} - \mathcal{A}_{t_j}\big)\mathcal{B}_{t_j}\mathrm{M}[\psi_j]f\big)(t) \big\|_X \Big)^p \, dt \leq$$

$$2^{p-1} \int_0^1 \sum_{j=1}^{N} \big\| \big(\mathrm{M}[\phi_j]\big(\mathcal{A} - \mathcal{A}_{t_j}\big)\mathcal{B}_{t_j}\mathrm{M}[\psi_j]f\big)(t) \big\|^p_X \, dt =$$

$$2^{p-1} \int_0^1 \sum_{j=1}^{N} \big\| \phi_j(t) \cdot \big(\big(\mathcal{A} - \mathcal{A}_{t_j}\big)\mathcal{B}_{t_j}\mathrm{M}[\psi_j]f\big)(t) \big\|^p_X \, dt =$$

$$2^{p-1} \sum_{j=1}^{N} \int_0^1 \phi_j(t)^p \big\| (A_t - A_{t_j})[(\mathcal{B}_{t_j}\mathrm{M}[\psi_j]f)(t)] \big\|^p_X \, dt \leq$$

$$2^{p-1} \sum_{j=1}^{N} \int_0^1 \mathbb{1}_{U_j}(t)^p \|A_t - A_{t_j}\|^p_{\mathscr{L}(D,X)} \|(\mathcal{B}_{t_j}\mathrm{M}[\psi_j]f)(t)\|^p_D \, dt \leq$$

$$2^{p-1}(\omega(\delta))^p \sum_{j=1}^{N} \int_0^1 \|(\mathcal{B}_{t_j}\mathrm{M}[\psi_j]f)(t)\|^p_D \, dt =$$

$$2^{p-1}(\omega(\delta))^p \sum_{j=1}^{N} \|\mathcal{B}_{t_j}\mathrm{M}[\psi_j]f\|^p_{L_p(\mathbb{T},D)} \leq$$

$$2^{p-1}(\omega(\delta))^p \sum_{j=1}^{N} \|\mathcal{B}_{t_j}\mathrm{M}[\psi_j]f\|^p_{\mathcal{W}_1(\mathbb{T})} \leq$$

$$2^{p-1}(\omega(\delta))^p \sum_{j=1}^{N} \|\mathcal{B}_{t_j}\|^p_{\mathscr{L}(\mathcal{W}_0[0,1],\mathcal{W}_1(\mathbb{T}))} \|\mathrm{M}[\psi_j]f\|^p_{\mathcal{W}_0[0,1]} \leq$$

$$2^{p-1}c_{\mathcal{B}}{}^p(\omega(\delta))^p\sum_{j=1}^N\|\mathrm{M}[\psi_j]f\|^p_{\mathcal{W}_0[0,1]}=$$

$$(2c_{\mathcal{B}}\omega(\delta))^p/2\sum_{j=1}^N\int_0^1\|(\mathrm{M}[\psi_j]f)(t)\|^p_X\,dt\leq$$

$$(2c_{\mathcal{B}}\omega(\delta))^p/2\int_0^1\Big(\sum_{j=1}^N\|(\mathrm{M}[\psi_j]f)(t)\|_X\Big)^p\,dt=$$

$$(2c_{\mathcal{B}}\omega(\delta))^p/2\int_0^1\Big(\sum_{j=1}^N\psi_j(t)\Big)^p\|f(t)\|^p_X\,dt=$$

$$(2c_{\mathcal{B}}\omega(\delta))^p/2\int_0^1\|f(t)\|^p_X\,dt=(2c_{\mathcal{B}}\omega(\delta))^p/2\,\|f\|^p_{\mathcal{W}_0[0,1]}.$$

We conclude that $\|\mathcal{S}\|_{\mathscr{L}(\mathcal{W}_0[0,1])}\leq 2^{1-p}c_{\mathcal{B}}\omega(\delta)\omega(\delta)\|f\|_{\mathcal{W}_0[0,1]}\overset{\delta\to 0}{\longrightarrow}0$. From now on we assume that $\delta$ is small enough such that $\|\mathcal{S}\|_{\mathscr{L}(\mathcal{W}_0[0,1])}<1$. Thus then $\mathrm{Id}_{\mathcal{W}_0[0,1]}+\mathcal{S}\in\mathscr{L}(\mathcal{W}_0[0,1])$ is invertible.

Since $\mathcal{L}_{\mathcal{W}_1(\mathbb{T})}\mathcal{B}=\mathrm{Id}_{\mathcal{W}_0[0,1]}+\mathcal{S}+\mathcal{K}_1+\mathcal{K}_2$ we obtain

$$\mathcal{L}_{\mathcal{W}_1(\mathbb{T})}\mathcal{B}(\mathrm{Id}_{\mathcal{W}_0[0,1]}+\mathcal{S})^{-1}=\mathrm{Id}_{\mathcal{W}_0[0,1]}+(\mathcal{K}_1+\mathcal{K}_2)(\mathrm{Id}_{\mathcal{W}_0[0,1]}+\mathcal{S})^{-1}.$$

Clearly, $(\mathcal{K}_1+\mathcal{K}_2)(\mathrm{Id}_{\mathcal{W}_0[0,1]}+\mathcal{S})^{-1}\in\mathscr{K}(\mathcal{W}_0[0,1])$. Thus

$$\mathcal{B}(\mathrm{Id}_{\mathcal{W}_0[0,1]}+\mathcal{S})^{-1}\in\mathscr{L}(\mathcal{W}_0[0,1],\mathcal{W}_1(\mathbb{T}))$$

is a right regularizer.

For the construction of a left regularizer we first remark that

$$\mathcal{B}_{t_j}\in\mathscr{L}(\mathcal{W}_0[0,1],\mathcal{W}_1(\mathbb{T}))$$

together with Fact 3.1.5 also yields $\mathcal{B}_{t_j}\in\mathscr{K}(\mathcal{W}_1(\mathbb{T}))$ for each $j=1,\ldots,N$.

Analogously as above we obtain $(\mathcal{A}-\mathcal{A}_{t_j})\mathrm{M}[\widetilde{\psi}_j]\in\mathscr{L}(\mathcal{W}_1(\mathbb{T}),\mathcal{W}_0[0,1])$ and $\|(\mathcal{A}-\mathcal{A}_{t_j})\mathrm{M}[\widetilde{\psi}_j]\|_{\mathscr{L}(\mathcal{W}_1(\mathbb{T}),\mathcal{W}_0[0,1])}\leq\omega(\delta)$ for each $j=1,\ldots,N$. Therefore $\|\mathcal{B}_{t_j}(\mathcal{A}-\mathcal{A}_{t_j})\mathrm{M}[\widetilde{\psi}_j]\|_{\mathscr{L}(\mathcal{W}_1(\mathbb{T}))}\leq c_{\mathcal{B}}\omega(\delta)\overset{\delta\to 0}{\longrightarrow}0$ and thus we can again assume that $\delta$ is small enough such that $\mathcal{B}_{t_j}(\mathcal{A}-\mathcal{A}_{t_j})\mathrm{M}[\widetilde{\psi}_j]$ is a contraction in $\mathcal{W}_1(\mathbb{T})$. Hence then $\mathrm{Id}_{\mathcal{W}_1(\mathbb{T})}+\mathcal{B}_{t_j}(\mathcal{A}-\mathcal{A}_{t_j})\mathrm{M}[\widetilde{\psi}_j]$ is invertible and we set $\mathcal{S}_j:=\big(\mathrm{Id}_{\mathcal{W}_1(\mathbb{T})}+\mathcal{B}_{t_j}(\mathcal{A}-\mathcal{A}_{t_j})\mathrm{M}[\widetilde{\psi}_j]\big)^{-1}\in\mathscr{L}(\mathcal{W}_1(\mathbb{T}))$.

A simple calculation now shows

$$\Big(\sum_{j=1}^N\phi_j\mathcal{S}_j\mathcal{B}_{t_j}\mathrm{M}[\psi_j]\Big)\mathcal{L}_{\mathcal{W}_1(\mathbb{T})}=$$

$$\mathrm{Id}_{\mathcal{W}_1(\mathbb{T})} + \rho \sum_{j=1}^N \mathrm{M}[\phi_j]\mathcal{S}_j\mathcal{B}_{t_j}\mathrm{M}[\psi_j] - \sum_{j=1}^N \mathrm{M}[\phi_j]\mathcal{S}_j\mathcal{B}_{t_j}\mathrm{M}[\partial\psi_j].$$

As above we obtain that the second and third summand are compact operators on $\mathcal{W}_1(\mathbb{T})$ since $\mathcal{B}_{t_j} \in \mathscr{K}(\mathcal{W}_1(\mathbb{T}))$. Thus $\sum_{j=1}^N \mathrm{M}[\phi_j]\mathcal{S}_j\mathcal{B}_{t_j}\mathrm{M}[\psi_j]$ is a left regularizer.

This proves the theorem. □

### 3.1.7 Corollary

Let ($A$-iv) and ($A$-v) hold.
Then $\mathcal{L}_{\mathcal{W}_1(\mathbb{T})} - z \in \mathscr{L}(\mathcal{W}_1(\mathbb{T}), \mathcal{W}_0[0,1])$ is a Fredholm operator for each $z \in \mathbb{C}$.

*Proof.*
Since by Fact 3.1.5 $\mathcal{W}_1(\mathbb{T}) \hookrightarrow\!\!\!\rightarrow \mathcal{W}_0[0,1]$ we conclude that $\mathcal{L}_{\mathcal{W}_1(\mathbb{T})} - z$ is a compact perturbation of the Fredholm operator $\mathcal{L}_{\mathcal{W}_1(\mathbb{T})}$. Thus [Sch73] Theorem 3.1 yields the statement. □

A direct calculation yields the following relation which will allow us to extend the Fredholm result to all operators $\mathcal{L}_{\mathcal{W}_{1,z}[0,1]}$.

### 3.1.8 Fact

Let $z \in \mathbb{C}\backslash\{0\}$.
Then $\mathcal{L}_{\mathcal{W}_{1,\xi}[0,1]} = \mathrm{M}[\xi_{(z)}^{(\cdot)}](\mathcal{L}_{\mathcal{W}_1(\mathbb{T})} + \log_{(z)} \xi)\mathrm{M}[\xi_{(z)}^{(-\cdot)}]$ for all $\xi \in \mathbb{B}_z$.

### 3.1.9 Corollary

Let ($A$-iv) and ($A$-v) hold.
Then $\mathcal{L}_{\mathcal{W}_{1,z}[0,1]} \in \mathscr{L}(\mathcal{W}_{1,z}[0,1], \mathcal{W}_0[0,1])$ is a Fredholm operator for each $z \in \mathbb{C}$.

*Proof.*
Let $z \in \mathbb{C}\backslash\{0\}$. Fact 3.1.8 particularly yields

$$\mathcal{L}_{\mathcal{W}_{1,z}[0,1]} = \mathrm{M}[z_{(z)}^{(\cdot)}](\mathcal{L}_{\mathcal{W}_1(\mathbb{T})} + \log_{(z)} z)\mathrm{M}[z_{(z)}^{(-\cdot)}].$$

By Corollary 3.1.7 $\mathcal{L}_{\mathcal{W}_1(\mathbb{T})} + \log_{(z)} z$ is a Fredholm operator and Fact 1.8.3 yields that $\mathrm{M}[z_{(z)}^{(-\cdot)}] \in \mathscr{L}(\mathcal{W}_{1,z}[0,1], \mathcal{W}_1(\mathbb{T}))$ and $\mathrm{M}[z_{(z)}^{(\cdot)}] \in \mathscr{L}(\mathcal{W}_0[0,1])$ are isomorphisms[9]. We obtain that $\mathcal{L}_{\mathcal{W}_{1,z}[0,1]}$ is a Fredholm operator. □

### 3.1.10 Remark

Indepently and during the development of this thesis, W. Arendt and P. Rabier proved similar results to this chapter, cf. [AR09].

---

[9] The inverses are $\mathrm{M}[z_{(z)}^{(\cdot)}] \in \mathscr{L}(\mathcal{W}_1(\mathbb{T}), \mathcal{W}_{1,z}[0,1])$ and $\mathrm{M}[z_{(z)}^{(-\cdot)}] \in \mathscr{L}(\mathcal{W}_0[0,1])$,resp..

# Chapter 4

# The Dual Framework

We will now introduce the "dual" objects to the objects defined in Chapter 2. Loosely speaking, all construction are carried out analogously by substituting $D$ by $D'$,$X$ by $X^*$ and $p$ by $q$.

## 4.1 $(A_t^*)_{t\in\mathbb{T}}$, $\mathcal{A}'$, $\mathcal{A}'_{t_0}$

We will use duality to define what we mean by solutions to equation (E). Therefore we assume from now on, that the following three conditions hold for the dual operators $A_t^* : X^* \supset D(A_t^*) \to X^*$.

($A^*$-i) There exists a normed space $(D', \|\cdot\|_{D'})$ such that $D(A_t^*) = D'$ for all $t \in \mathbb{R}$.

($A^*$-ii) $\|\cdot\|_{D'}$ is, uniformly in $t \in \mathbb{R}$, equivalent to all graph norms $\|\cdot\|_{A_t^*}$.

($A^*$-iii) $[t \mapsto A_t^*] \in C(\mathbb{T}, \mathscr{L}(D', X^*))$.

**4.1.1 Remark**
ince $A_t$ (for, say, $t := 0$) is closed, $(D', \|\cdot\|_{X^*})$ is dense in $X^*$ by [Kat66] Theorem III.5.29.

Periodicity of $(A_t)_{t\in\mathbb{R}}$ yields periodicity of $(A_t^*)_{t\in\mathbb{R}}$ and we will again use the notation $t \in \mathbb{T}$ when referring to indices.

Analogously as in Section 2.2 and Section 2.3, resp., we then define the closed, densely defined operators $\mathcal{A}' : L_q([0,1], X^*) \supset L_q([0,1], D') \to L_q([0,1], X^*)$ and $\mathcal{A}'_{t_0} : L_q([0,1], X^*) \supset L_q([0,1], D') \to L_q([0,1], X^*)$ for (a fixed) $t_0 \in \mathbb{T}$.

## 4.2 $\mathcal{W}_1'[a,b]$, $\mathcal{W}_{1,z}'[0,1]$, $\mathcal{W}_0'[a,b]$, $\mathcal{W}_{-1,z}'[0,1]$, $\mathcal{W}_1'(\mathbb{T})$

Analogously to Section 2.4, we define, again for all $z \in \mathbb{C}\backslash\{0\}$ and $a, b \in \mathbb{R}$ with $a < b$, the Banach spaces

$$\mathcal{W}_1'[a,b] := L_q([a,b], D') \cap W_q^1([a,b], X^*),$$

endowed with the norm $\|f\|_{\mathcal{W}_1'[a,b]} := \left(\|f\|_{L_q([a,b],D')}^q + \|\partial f\|_{L_q([a,b],X^*)}^q\right)^{1/q}$,

$$\mathcal{W}_{1,z}'[0,1] := L_q([0,1], D') \cap W_q^1([0,1], X^*)_z,$$

endowed with the norm $\|f\|_{\mathcal{W}_{1,z}'[0,1]} := \|f\|_{\mathcal{W}_1'[0,1]}$,

$$\mathcal{W}_0'[a,b] := L_q([a,b], X^*), \text{ and}$$

$$\mathcal{W}_{-1,z}'[0,1] := \left(\mathcal{W}_{1,1/z}'[0,1]\right)^* \cong \left(L_q([0,1], D')\right)^* + \left(W_q^1([0,1], X^*)_{1/z}\right)^*.$$

Again, we set

$$\mathcal{W}_1'(\mathbb{T}) := \mathcal{W}_{1,1}'[0,1]$$

and we state the results analogous to Fact 2.4.1 and Fact 2.4.2.

**4.2.1 Fact**
$\delta_t \in \mathscr{L}(\mathcal{W}_1'[0,1], X^*)$ for all $t \in [0,1]$.

**4.2.2 Fact**
$C^\infty([0,1], D')_z$ is dense in $\mathcal{W}_{1,z}'[0,1]$.

**4.2.3 Definition and Fact** (*Duality between* $\mathcal{W}_0[0,1]$ *and* $\mathcal{W}_0'[0,1]$)
Since $X$ is reflexive, by [Edw65] Theorem 8.20.5 $\left(\mathcal{W}_0[0,1]\right)^*$ can be identified with $\mathcal{W}_0'[0,1]$ by the usual isomorphism $\mathcal{W}_0'[0,1] \ni f' \mapsto \langle f', \cdot \rangle_{\mathcal{W}} \in \left(\mathcal{W}_0[0,1]\right)^*$, where $\langle f', f \rangle_{\mathcal{W}} := \int_0^1 \langle f'(t), f(t) \rangle_X \, dt$ for all $f' \in \mathcal{W}_0'[0,1]$ and $f \in \mathcal{W}_0[0,1]$. Analogously, $\mathcal{W}_0[0,1] \cong \left(\mathcal{W}_0'[0,1]\right)^*$ by the isomorphism $f \mapsto \langle \cdot, f \rangle_{\mathcal{W}}$. Furthermore, both $\mathcal{W}_0[0,1]$ and $\mathcal{W}_0'[0,1]$ are reflexive and as usual the canonical embedding into their biduals is compatible with the identification $\langle \cdot, \cdot \rangle_{\mathcal{W}}$. We will use these identifications without further notice.

## 4.3 Extension and Duality of $\mathcal{L}$ and $\mathcal{L}'$

**4.3.1 Motivation**
In this section we will introduce extensions of the operators $\mathcal{L}_{\mathcal{W}_{1,z}[0,1]}$ and $\mathcal{L}'_{\mathcal{W}_{1,z}'[0,1]}$ using duality. We will explain the basic idea behind this construction in the following (general) situation.

If $a : x \supset d \to x$ is a closed and densely defined operator on a reflexive Banach space $x$, then there are the following two operators associated with $a$ w. r. t. duality.

First, there is the "usual" dual operator $a^* : x^* \supset d' \to x^*$ (as e. g. defined in [Kat66] § III.5.5), which then is also closed and densely defined by [Kat66] Theorem III.5.29.

On the other hand, $a \in \mathscr{L}([d], x)$, where $[d]$ denotes the Banach space induced by the graph norm of $a$ on $d$. As a bounded operator $a$ has then a dual operator in $\mathscr{L}(x^*, [d]^*)$ (in the sense of [Kat66] § I.3.6), which we (here) denote by $a^{*\sim}$.

If we apply the second method to $a^* : x^* \supset d' \to x^*$, we obtain $a^\sim := (a^*)^{*\sim} \in \mathscr{L}(x^{**}, [d']^*)$. After the natural identification of $x^{**}$ with $x$, we obtain $a^\sim \in \mathscr{L}(x, [d']^*)$ and a direct calculation shows that $a^\sim$ is the (unique $[d']^*$-valued) extension of $a \in \mathscr{L}([d], x)$ to $x$. △

During this section let $z \in \mathbb{C}\backslash\{0\}$. We remark that the definition of the following operators, in particular the "extended ones", depends on $z$. E. g., if in the above motivation $a = \partial$ with $x = L_p([0,1], \mathbb{C})$ and $d = W_p^1([0,1], \mathbb{C})_z$, then $a^\sim \mathbb{1} = (1 - 1/z)\delta_0$ (in $[d']^*$)). However, for the sake of readability we will mostly omit the use of a corresponding index in this section.

We remark that $\mathcal{A} : \mathcal{W}_0[0,1] \supset L_p([0,1], D) \to \mathcal{W}_0[0,1]$ is the adjoint operator of $\mathcal{A}' : \mathcal{W}_0'[0,1] \supset L_q([0,1], D') \to \mathcal{W}_0'[0,1]$ and vice versa.

We have already mentioned that the graph norm of $\mathcal{A}$ is equivalent to the norm of $L_p([0,1], D)$. Thus $\mathcal{A} \in \mathscr{L}(L_p([0,1], D), \mathcal{W}_0[0,1])$. Analogously, we obtain $\mathcal{A}' \in \mathscr{L}(L_q([0,1], D'), \mathcal{W}_0'[0,1])$.

For both operators there is a unique extension to

$\mathcal{A}^\sim \in \mathscr{L}\big(\mathcal{W}_0[0,1], \big(L_q([0,1], D')\big)^*\big)$ and

$\mathcal{A}'^\sim \in \mathscr{L}(\mathcal{W}_0'[0,1], (L_p([0,1], D))^*)$, resp..

$\mathcal{A}^\sim$ coincides with the dual operator of $\mathcal{A}' \in \mathscr{L}(L_q([0,1], D'), \mathcal{W}_0'[0,1])$ and $\mathcal{A}'^\sim$ coincides with the dual operator of $\mathcal{A} \in \mathscr{L}(L_p([0,1], D), \mathcal{W}_0[0,1])$.

For the sake of completeness we remark that analogous results hold for $\mathcal{A}_{t_0}$ and $\mathcal{A}_{t_0}'$, resp. and we define $\mathcal{A}_{t_0}^\sim \in \mathscr{L}\big(\mathcal{W}_0[0,1], \big(L_q([0,1], D')\big)^*\big)$ and $\mathcal{A}_{t_0}'^\sim \in \mathscr{L}(\mathcal{W}_0'[0,1], (L_p([0,1], D))^*)$ analogously.

Furthermore, as usual by partial integration $\partial : \mathcal{W}_0[0,1] \supset W_p^1([0,1], X)_z \to$

$\mathcal{W}_0[0,1]$ and $-\partial : \mathcal{W}_0'[0,1] \supset W_q^1([0,1],X^*)_{1/z} \to \mathcal{W}_0'[0,1]$ are closed and densely defined operators that are mutually adjoint. Their graph norms are equivalent to the given norms on their domains.

Again, the dual operator of $-\partial \in \mathscr{L}(W_q^1([0,1],X^*)_{1/z}, \mathcal{W}_0'[0,1])$ is an extension of $\partial \in \mathscr{L}(W_p^1([0,1],X)_z, \mathcal{W}_0[0,1])$ to

$$\partial^\sim \in \mathscr{L}\big(\mathcal{W}_0[0,1], \big(W_q^1([0,1],X^*)_{1/z}\big)^*\big)$$

and the dual operator of $\partial \in \mathscr{L}(W_p^1([0,1],X)_z, \mathcal{W}_0[0,1])$ is an extension of $-\partial \in \mathscr{L}(W_q^1([0,1],X^*)_{1/z}, \mathcal{W}_0'[0,1])$ to

$$-\partial^\sim \in \mathscr{L}\big(\mathcal{W}_0'[0,1], \big(W_p^1([0,1],X)_z\big)^*\big).$$

The following diagrams on the next page illustrate the situation. To avoid confusion, we explicitly state that the symbol $\cap$ denotes the intersection.

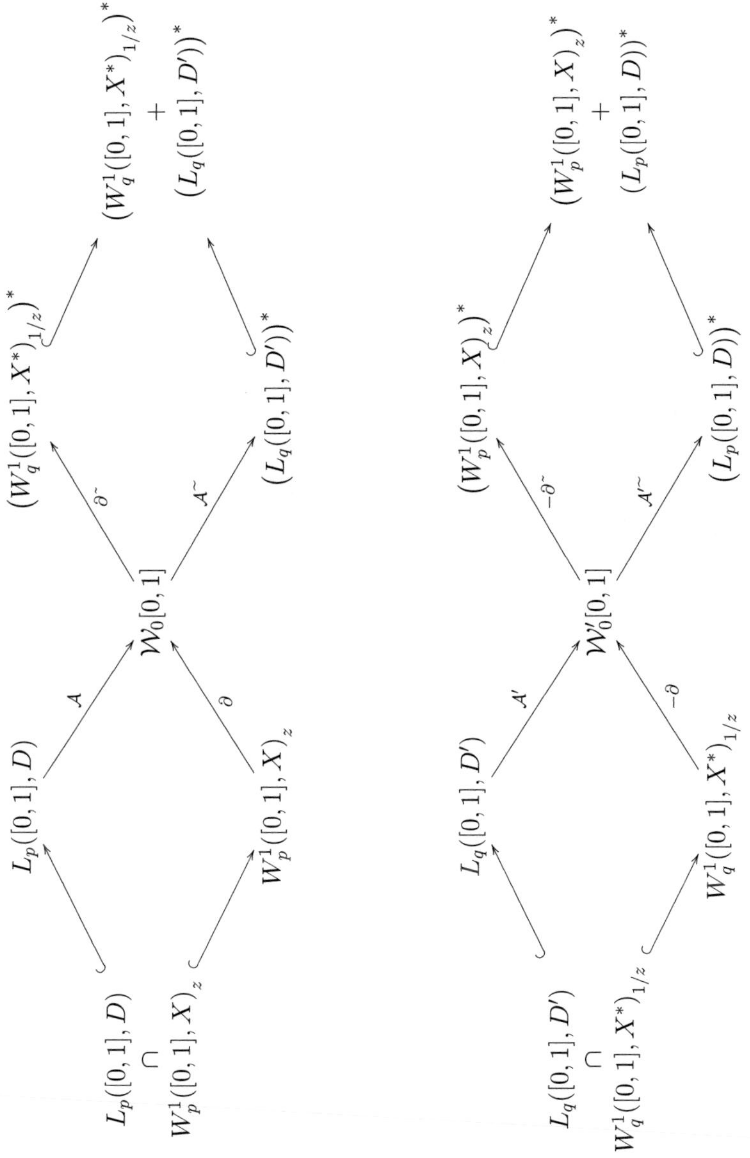
$L_p([0,1],D) \cap W_p^1([0,1],X)_z$
$L_p([0,1],D)$
$W_p^1([0,1],X)_z$
$\mathcal{A}$
$\partial$
$\mathcal{W}_0[0,1]$
$\partial^\sim$
$\mathcal{A}^\sim$
$(W_q^1([0,1],X^*)_{1/z})^*$
$(L_q([0,1],D'))^*$
$(W_q^1([0,1],X^*)_{1/z})^* + (L_q([0,1],D'))^*$
$L_q([0,1],D') \cap W_q^1([0,1],X^*)_{1/z}$
$L_q([0,1],D')$
$W_q^1([0,1],X^*)_{1/z}$
$\mathcal{A}'$
$-\partial$
$\mathcal{W}_0'[0,1]$
$-\partial^\sim$
$\mathcal{A}'^\sim$
$(W_p^1([0,1],X)_z)^*$
$(L_p([0,1],D))^*$
$(W_p^1([0,1],X)_z)^* + (L_p([0,1],D))^*$

We recall that in Chapter 3 we defined

$$\mathcal{L}_{\mathcal{W}_{1,z}[0,1]} := \partial + \mathcal{A} \in \mathscr{L}(\mathcal{W}_{1,z}[0,1], \mathcal{W}_0[0,1]),$$

and we now set

$$\mathcal{L}_z^{\sim} := \partial^{\sim} + \mathcal{A}^{\sim} \in \mathscr{L}(\mathcal{W}_0[0,1], \mathcal{W}'_{-1,z}[0,1]),$$

$$\mathcal{L}'_{\mathcal{W}'_{1,1/z}[0,1]} := -\partial + \mathcal{A}' \in \mathscr{L}(\mathcal{W}'_{1,1/z}[0,1], \mathcal{W}'_0[0,1]) \text{ and}$$

$$\mathcal{L}'^{\sim}_{1/z} := -\partial^{\sim} + \mathcal{A}'^{\sim} \in \mathscr{L}(\mathcal{W}'_0[0,1], \mathcal{W}_{-1,1/z}[0,1]).$$

We remark that throughout this thesis, formally $\mathcal{L} := \partial + \mathcal{A}$ and $\mathcal{L}' := -\partial + \mathcal{A}'$ will hold and subscripts will be used for concrete realizations.

It follows that $\mathcal{L}_z^{\sim}$ is the unique extension $\mathcal{L}_{\mathcal{W}_{1,z}[0,1]}$ and coincides with the dual operator of $\mathcal{L}'_{\mathcal{W}'_{1,1/z}[0,1]}$. Analogously, $\mathcal{L}'^{\sim}_{1/z}$ is the unique extension $\mathcal{L}'_{\mathcal{W}'_{1,1/z}[0,1]}$ and coincides with the dual operator of $\mathcal{L}_{\mathcal{W}_{1,z}[0,1]}$.

For the convenience of the reader we redraw the diagrams using the introduced notations.

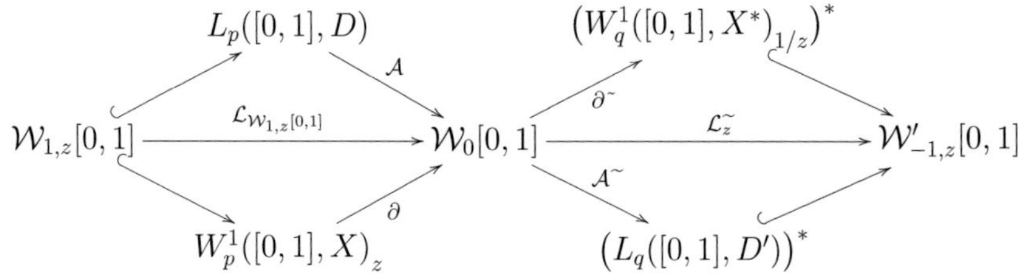


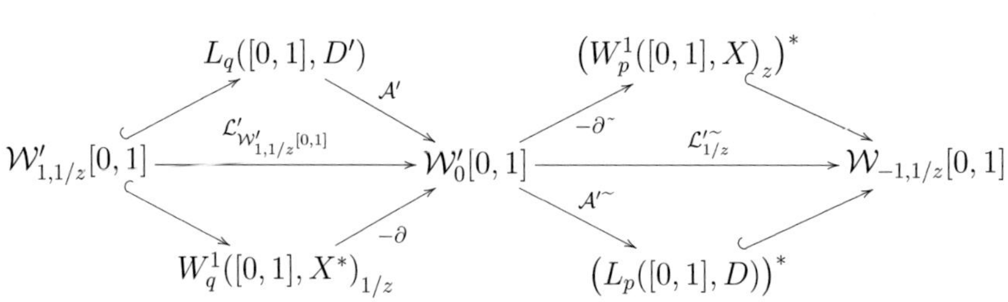

## 4.4 The Operators $\mathcal{L}$ and $\mathcal{L}'$ on the Real Line – Part 1

The following statements are direct consequences of the definitions and the underlying spaces, cf. Section 1.3.

**4.4.1 Definition and Remark**

We set

$\mathcal{W}_{1,\mathrm{loc}}(\mathbb{R}) := L_{p,\mathrm{loc}}(\mathbb{R}, D) \cap W^1_{p,\mathrm{loc}}(\mathbb{R}, X)$,

$\mathcal{W}_{0,\mathrm{loc}}(\mathbb{R}) := L_{p,\mathrm{loc}}(\mathbb{R}, X)$,

$\mathcal{W}'_{1,\mathrm{loc}}(\mathbb{R}) := L_{q,\mathrm{loc}}(\mathbb{R}, D') \cap W^1_{q,\mathrm{loc}}(\mathbb{R}, X^*)$ and

$\mathcal{W}'_{0,\mathrm{loc}}(\mathbb{R}) := L_{q,\mathrm{loc}}(\mathbb{R}, X^*)$.

Obviously, (an equivalence class of) a function $f : \mathbb{R} \to X$ belongs to $\mathcal{W}_{1,\mathrm{loc}}(\mathbb{R})$ or $\mathcal{W}_{0,\mathrm{loc}}(\mathbb{R})$ iff for all $a, b \in \mathbb{R}$ with $a < b$ the restriction $f_{|[a,b]}$ belongs to $\mathcal{W}_1[a,b]$ or $\mathcal{W}_0[a,b]$, resp..

Analogously, (an equivalence class of) a function $f' : \mathbb{R} \to X^*$ belongs to $\mathcal{W}'_{1,\mathrm{loc}}(\mathbb{R})$ or $\mathcal{W}'_{0,\mathrm{loc}}(\mathbb{R})$ iff for all $a, b \in \mathbb{R}$ with $a < b$ the restriction $f'_{|[a,b]}$ belongs to $\mathcal{W}'_1[a,b]$ or $\mathcal{W}'_0[a,b]$, resp..

We endow $\mathcal{W}_{1,\mathrm{loc}}(\mathbb{R})$, $\mathcal{W}_{0,\mathrm{loc}}(\mathbb{R})$, $\mathcal{W}'_{1,\mathrm{loc}}(\mathbb{R})$ and $\mathcal{W}'_{0,\mathrm{loc}}(\mathbb{R})$ with the topologies induced by the seminorms

$\{\|\cdot\|_{\mathcal{W}_1[-k,k]}\}_{k\in\mathbb{N}}$, $\{\|\cdot\|_{\mathcal{W}_0[-k,k]}\}_{k\in\mathbb{N}}$, $\{\|\cdot\|_{\mathcal{W}'_1[-k,k]}\}_{k\in\mathbb{N}}$ and

$\{\|\cdot\|_{\mathcal{W}'_0[-k,k]}\}_{k\in\mathbb{N}}$, resp..

Then $\mathcal{W}_{1,\mathrm{loc}}(\mathbb{R})$ (and analogously $\mathcal{W}_{0,\mathrm{loc}}(\mathbb{R})$, $\mathcal{W}'_{1,\mathrm{loc}}(\mathbb{R})$ and $\mathcal{W}'_{0,\mathrm{loc}}(\mathbb{R})$) is a Fréchet space (since it coincides with the projective limit of the Banach spaces $\{\mathcal{W}_1[-k,k]\}_{k\in\mathbb{N}}$, cf. [FW68] § 6.2).

Furthermore, we easily obtain:

**4.4.2 Fact**

$f \in \mathcal{W}_{1,\mathrm{loc}}(\mathbb{R})$ iff $f \in \mathcal{W}_1[k, k+1]$ and $f_{|[k-1,k]}(k) = f_{|[k,k+1]}(k)$ for all $k \in \mathbb{Z}$.

$f \in \mathcal{W}_{0,\mathrm{loc}}(\mathbb{R})$ iff $f \in \mathcal{W}_0[k, k+1]$ for all $k \in \mathbb{Z}$.

$f' \in \mathcal{W}'_{1,\mathrm{loc}}(\mathbb{R})$ iff $f' \in \mathcal{W}'_1[k, k+1]$ and $f'_{|[k-1,k]}(k) = f'_{|[k,k+1]}(k)$ for all $k \in \mathbb{Z}$.

$f' \in \mathcal{W}'_{0,\mathrm{loc}}(\mathbb{R})$ iff $f' \in \mathcal{W}'_0[k, k+1]$ for all $k \in \mathbb{Z}$.

We now define the operators $\mathcal{L}$ and $\mathcal{L}'$ on the real line.

**4.4.3 Definition and Fact**

$\mathcal{L}_{\mathcal{W}_{1,\mathrm{loc}}(\mathbb{R})} := [f \mapsto (\partial + \mathcal{A})f] \in \mathscr{L}(\mathcal{W}_{1,\mathrm{loc}}(\mathbb{R}), \mathcal{W}_{0,\mathrm{loc}}(\mathbb{R}))$.

$\mathcal{L}'_{\mathcal{W}'_{1,\mathrm{loc}}(\mathbb{R})} := [f' \mapsto (-\partial + \mathcal{A}')f'] \in \mathscr{L}(\mathcal{W}'_{1,\mathrm{loc}}(\mathbb{R}), \mathcal{W}'_{0,\mathrm{loc}}(\mathbb{R}))$.

**4.4.4 Remark**
Obviously, $C_c^\infty(\mathbb{R}, D') \subset \mathcal{W}'_{1,\mathrm{loc}}(\mathbb{R})$ and $\mathcal{L}'_{\mathcal{W}'_{1,\mathrm{loc}}(\mathbb{R})}(C_c^\infty(\mathbb{R}, D')) \subset C_c(\mathbb{R}, X^*)$.

Combining Section 1.4 with Fact 4.4.2, a direct calculation yields:

**4.4.5 Fact**
Let $f \in \mathcal{W}_{1,z}[0,1]$.
Then $\mathrm{E}_z f \in \mathcal{W}_{1,\mathrm{loc}}(\mathbb{R})$ and $\mathcal{L}_{\mathcal{W}_{1,\mathrm{loc}}(\mathbb{R})}(\mathrm{E}_z f) = \mathrm{E}_z(\mathcal{L}_{\mathcal{W}_{1,z}[0,1]} f)$ (in $\mathcal{W}_{0,\mathrm{loc}}(\mathbb{R})$).

**4.4.6 Fact**
Let $f' \in \mathcal{W}'_{1,z}[0,1]$.
Then $\mathrm{E}_z f' \in \mathcal{W}'_{1,\mathrm{loc}}(\mathbb{R})$ and $\mathcal{L}'_{\mathcal{W}'_{1,\mathrm{loc}}(\mathbb{R})}(\mathrm{E}_z f') = \mathrm{E}_z(\mathcal{L}'_{\mathcal{W}'_{1,z}[0,1]} f')$ (in $\mathcal{W}'_{0,\mathrm{loc}}(\mathbb{R})$).

## 4.5 Solutions – Part 1

Using the introduced notation, equation (E) formally reads $\mathcal{L}u = 0$. We will now give the precise definition of solutions, using duality.

**4.5.1 Preparation and Definition**
For all $f' \in C_c(\mathbb{R}, X^*)$ and $f \in \mathcal{W}_{0,\mathrm{loc}}(\mathbb{R})$ clearly $[t \mapsto \langle f'(t), f(t)\rangle_X] \in L_1(\mathbb{R}, \mathbb{C})$ and we set $\langle f', f\rangle := \int_{\mathbb{R}} \langle f'(t), f(t)\rangle_X \, dt \in \mathbb{C}$.
In particular, by Remark 4.4.4 $\langle \mathcal{L}'_{\mathcal{W}'_{1,\mathrm{loc}}(\mathbb{R})}\phi', f\rangle$ is well-defined for all $\phi' \in C_c^\infty(\mathbb{R}, D')$ and $f \in \mathcal{W}_{0,\mathrm{loc}}(\mathbb{R})$.

**4.5.2 Definition** (*Solutions*)
$u \in \mathcal{W}_{0,\mathrm{loc}}(\mathbb{R})$ is called a *solution (to equation* (E)*)* if $\langle \mathcal{L}'_{\mathcal{W}'_{1,\mathrm{loc}}(\mathbb{R})}\phi', u\rangle = 0$ for all $\phi' \in C_c^\infty(\mathbb{R}, D')$.

We end this section with two propositions that mostly will play a technical role but also indicate the connection between the "weak formulation" used in the above definition and a more classical concept.

**4.5.3 Proposition**
Let $u \in \mathcal{W}_{1,\mathrm{loc}}(\mathbb{R})$.
Then $u$ is a solution iff $\mathcal{L}_{\mathcal{W}_{1,\mathrm{loc}}(\mathbb{R})}u = 0$.

*Proof.*
Analogously as in Section 4.3 partial integration yields

$$\langle \mathcal{L}'_{\mathcal{W}'_{1,\mathrm{loc}}(\mathbb{R})}\phi', u\rangle = \langle \phi', \mathcal{L}_{\mathcal{W}_{1,\mathrm{loc}}(\mathbb{R})}u\rangle \text{ for all } \phi' \in C_c^\infty(\mathbb{R}, D').$$

Thus if $\mathcal{L}_{\mathcal{W}_{1,\mathrm{loc}}(\mathbb{R})}u = 0$ then we obtain directly that $u$ is a solution. Conversely, if $u$ is a solution, then density of the restrictions of $C_c^\infty(\mathbb{R}, D')$ in $\mathcal{W}'_0[a,b]$ yields $\mathcal{L}_{\mathcal{W}_{1,\mathrm{loc}}(\mathbb{R})}u = 0$ almost everywhere on $[a,b]$ for all $a, b \in \mathbb{R}$ with $a < b$, hence $\mathcal{L}_{\mathcal{W}_{1,\mathrm{loc}}(\mathbb{R})}u = 0$. □

### 4.5.4 Proposition

Let $u$ be a solution. Furthermore, assume that $u$ is $z$-quasiperiodic for some $z \in \mathbb{C}\backslash\{0\}$.
Then $\mathcal{L}_z^{\sim}(u_{|[0,1]}) = 0$.

*Proof.*
Let $\psi' \in C^\infty([0,1], D')_{1/z}$. We will show that $\langle \mathcal{L}_z^{\sim}(u_{|[0,1]}), \psi'\rangle_{\mathcal{W}'_{1,1/z}[0,1]} = 0$. Then the assertion follows from Fact 4.2.2.

Let $\alpha \in C_c^\infty(\mathbb{R}, \mathbb{C})$ with $\operatorname{supp}\alpha \subset [-1,2]$ and $\alpha = \mathbb{1}$ on $[0,1]$. We set $\tilde{\alpha} := \alpha$ on $(-\infty, 1]$, $\tilde{\alpha} := \mathbb{1}$ on $(1,2]$ and $\tilde{\alpha} := \alpha(\cdot - 1)$ on $(2,\infty)$. Then $\tilde{\alpha} \in C_c^\infty(\mathbb{R}, \mathbb{C})$ and $\phi' := \mathrm{M}[\alpha](\mathrm{E}_{1/z}\psi'), \tilde{\phi}' := \mathrm{M}[\tilde{\alpha}](\mathrm{E}_{1/z}\psi') \in C_c^\infty(\mathbb{R}, D')$.

Quasiperiodicity of $u$ and $\psi'$ in combination with the results of Section 4.3 and Fact 4.4.6 yields

$$\langle \mathcal{L}_z^{\sim}(u_{|[0,1]}), \psi'\rangle_{\mathcal{W}'_{1,1/z}[0,1]} = \langle u_{|[0,1]}, \mathcal{L}'_{\mathcal{W}'_{1,1/z}[0,1]}\psi'\rangle_{\mathcal{W}'_{1,1/z}[0,1]} =$$
$$\int_0^1 \langle (\mathcal{L}'_{\mathcal{W}'_{1,\mathrm{loc}}(\mathbb{R})}\psi')(t), u(t)\rangle_X \, dt = \int_1^2 \langle (\mathcal{L}'_{\mathcal{W}'_{1,\mathrm{loc}}(\mathbb{R})}(\mathrm{E}_{1/z}\psi'))(t), u(t)\rangle_X \, dt.$$

On the other hand, a direct calculation yields

$$\langle \mathcal{L}'_{\mathcal{W}'_{1,\mathrm{loc}}(\mathbb{R})}\tilde{\phi}', u\rangle = \langle \mathcal{L}'_{\mathcal{W}'_{1,\mathrm{loc}}(\mathbb{R})}\phi', u\rangle + \int_1^2 \langle (\mathcal{L}'_{\mathcal{W}'_{1,\mathrm{loc}}(\mathbb{R})}(\mathrm{E}_{1/z}\psi'))(t), u(t)\rangle_X \, dt.$$

By the assumption $\langle \mathcal{L}'_{\mathcal{W}'_{1,\mathrm{loc}}(\mathbb{R})}\phi', u\rangle = 0$ and $\langle \mathcal{L}'_{\mathcal{W}'_{1,\mathrm{loc}}(\mathbb{R})}\tilde{\phi}', u\rangle = 0$, hence

$$\int_1^2 \langle (\mathcal{L}'_{\mathcal{W}'_{1,\mathrm{loc}}(\mathbb{R})}(\mathrm{E}_{1/z}\psi'))(t), u(t)\rangle_X \, dt = 0.$$

This proves the proposition. □

## 4.6 Fredholm Property of $\mathcal{L}'$

For the sake of completeness we state the analogous result to Section 3. We remark that all statements easily follow from the fact that the family $(-A_t^*)_{t\in\mathbb{T}}$ fulfills the corresponding conditions.

**4.6.1 Theorem**
Assume that for (a fixed) $t_0 \in \mathbb{T}$ there exists $\rho \in \mathbb{R}$ such that $\rho + i\mathbb{R} \subset \rho(A_{t_0}^*)$ and $\{\, (|\lambda|+1)(A_{t_0}^* - \lambda)^{-1} : \lambda \in \rho+i\mathbb{R} \,\}$ is $(X^*, X^*)$-$R$-bounded. Furthermore, we define $\mathcal{L}'_{t_0} \in \mathscr{L}(\mathcal{W}'_1(\mathbb{T}), \mathcal{W}'_0[0,1])$ by $\mathcal{L}'_{t_0} := -\partial + \mathcal{A}'_{t_0} - \rho$. Then $\mathcal{L}'_{t_0}$ is invertible.

**4.6.2 Corollary**
Assume that for the family $(A_t^*)_{t\in\mathbb{T}}$ the following condition holds.

($A^*$-iv) There exists $\rho \in \mathbb{R}$ such that $\rho + i\mathbb{R} \subset \rho(A_t^*)$ and

$$\{\, (|\lambda| + 1)(A_t^* - \lambda)^{-1} : \lambda \in \rho + i\mathbb{R} \,\}$$

is uniformly $(X^*, X^*)$-$R$-bounded for all $t \in \mathbb{T}$.

Then $\{\, \mathcal{B}'_t : t \in \mathbb{T} \,\}$, where $\mathcal{B}'_t$ is the inverse of $\mathcal{L}'_t$ according to Theorem 4.6.1, is bounded (in $\mathscr{L}(\mathcal{W}'_0[0,1], \mathcal{W}'_1(\mathbb{T}))$).

**4.6.3 Fact**
Assume that

($A^*$-v) $D' \overset{c}{\hookrightarrow}\!\!\!\!\rightarrow X^*$

holds.
Then $\mathcal{W}'_1(\mathbb{T}) \overset{c}{\hookrightarrow}\!\!\!\!\rightarrow \mathcal{W}'_0[0,1]$.

**4.6.4 Theorem**
Let ($A^*$-iv) and ($A^*$-v) hold.
Then $\mathcal{L}'_{\mathcal{W}'_1(\mathbb{T})}$ is a Fredholm operator.

**4.6.5 Corollary**
Let ($A^*$-iv) and ($A^*$-v) hold.
Then $\mathcal{L}'_{\mathcal{W}'_1(\mathbb{T})} - z \in \mathscr{L}(\mathcal{W}'_1(\mathbb{T}), \mathcal{W}'_0[0,1])$ is a Fredholm operator for each $z \in \mathbb{C}$.

**4.6.6 Fact**
Let $z \in \mathbb{C}\backslash\{0\}$.
Then $\mathcal{L}'_{\mathcal{W}'_{1,\xi}[0,1]} = \mathrm{M}[\xi_{(z)}^{(\cdot)}](\mathcal{L}'_{\mathcal{W}'_1(\mathbb{T})} - \log_{(z)} \xi)\mathrm{M}[\xi_{(z)}^{(-\cdot)}]$ for all $\xi \in \mathbb{B}_z$.

**4.6.7 Corollary**
Let ($A^*$-iv) and ($A^*$-v) hold.
Then $\mathcal{L}'_{\mathcal{W}'_{1,z}[0,1]} \in \mathscr{L}(\mathcal{W}'_{1,z}[0,1], \mathcal{W}'_0[0,1])$ is a Fredholm operator for each $z \in \mathbb{C}$.

## 4.7 Equivalence and Relation of Conditions

**4.7.1 Proposition**
Condition ($A$-iv) is equivalent to condition ($A^*$-iv).

*Proof.*
We first note that $\rho(A_t^*) = \rho(A_t)$ and $(A_t^* - \lambda)^{-1} = ((A_t - \lambda)^{-1})^*$ for all $\lambda \in \rho(A_t^*)$ and all $t \in \mathbb{T}$. Then by [KKW06] Proposition 3.5 the uniform $(X, X)$-$R$-boundedness of $\{ (|\lambda| + 1)(A_t - \lambda)^{-1} : \lambda \in \rho + i\mathbb{R} \}$ for all $t \in \mathbb{T}$ yields the uniform $(X^*, X^*)$-$R$-boundedness of $\{ (|\lambda| + 1)(A_t^* - \lambda)^{-1} : \lambda \in \rho + i\mathbb{R} \}$, thus ($A$-iv) implies ($A^*$-iv). The converse direction directly follows from $(A_t^*)^* = A_t$ for all $t \in \mathbb{T}$. □

**Important Remark**
We will not always explicitly mention the above equivalence.

**4.7.2 Remark**
In particular, whenever in the following we require both the conditions ($A$-iv) and ($A^*$-iv) to hold, we can assume that they hold for the same $\rho \in \mathbb{R}$.

**4.7.3 Proposition**
Let ($A$-iv) hold.
Then condition ($A$-v) is equivalent to condition ($A^*$-v).

*Remarks on the proof.*
To avoid confusion we here use the symbol $^{*\sim}$ to denote the dual of a bounded operator, cf. Motivation 4.3.1.

*Proof.*
Assume that condition ($A$-v) holds and denote by $J : D \to X$ the compact embedding. Then by Schauder's theorem $J^{*\sim} \in \mathscr{K}(X^*, D^*)$. By ($A$-iv) $A_0 - \rho \in \mathscr{L}(D, X)$ is an isomorphism and hence $(A_0 - \rho)^{*\sim} \in \mathscr{L}(X^*, D^*)$ is also an isomorphism. Furthermore, by Proposition 4.7.1 $A_0^* - \rho \in \mathscr{L}(D', X^*)$ is an

isomorphism. An easy calculation now shows that the diagram

$$\begin{array}{ccc} X^* & \xrightarrow{J^{*\sim}} & D^* \\ {\scriptstyle A_0^*-\rho}\uparrow & & \downarrow{\scriptstyle ((A_0-\rho)^{*\sim})^{-1}} \\ D' & \xrightarrow[J']{} & X^* \end{array}$$

commutes, where $J'$ denotes the embedding of $D'$ into $X^*$. We conclude that $J'$ is compact and thus ($A^*$-v). Conversely, if we assume that ($A^*$-v) holds, i. e. $J' \in \mathscr{K}(D', X^*)$, then the diagram yields $J^{*\sim} \in \mathscr{K}(X^*, D^*)$ and again by Schauder's theorem we conclude that ($A$-v) holds. □

**4.7.4 Proposition**

Assume that there is $\rho \in \rho(A_t)$ for all $t \in \mathbb{R}$. Then the conditions ($A$-i) and ($A$-iii) and continuity of the embedding $D \hookrightarrow X$ imply condition ($A$-ii).

*Proof.*

Clearly ($A$-iii) yields that $c_1 := \sup_{t\in\mathbb{R}} \|A_t\|_{\mathscr{L}(D,X)} < \infty$ and by $D \hookrightarrow X$ there is $c_2 > 0$ such that $\|d\|_X \le c_2\|d\|_D$ for all $d \in D$. Thus

$$\|d\|_{A_t} \le \|A_t\|_{\mathscr{L}(D,X)}\|d\|_D + \|d\|_X \le \max\{c_1, c_2\}\|d\|_D$$

for all $d \in D$ and all $t \in \mathbb{R}$. On the other hand, by $[t \mapsto A_t - \rho] \in \mathscr{L}(D, X)$ we obtain $[t \mapsto (A_t - \rho)^{-1}] \in \mathscr{L}(X, D)$ (cf. [Cha85] Theorem 7.17) and thus $c_3 := \sup_{t\in\mathbb{R}} \|(A_t - \rho)^{-1}\|_{\mathscr{L}(X,D)} < \infty$. Therefore

$$\|d\|_D \le \|(A_t - \rho)^{-1}\|_{\mathscr{L}(X,D)}\|(A_t - \rho)d\|_X \le$$
$$c_3 \max\{1, |\rho|\}(\|A_t d\|_X + \|d\|_X) = c_3 \max\{1, |\rho|\}\|d\|_{A_t}$$

for all $d \in D$ and all $t \in \mathbb{R}$. □

As a direct consequence we obtain

**4.7.5 Corollary**

The conditions ($A$-i), ($A$-iii), ($A$-iv) and ($A$-v) imply condition ($A$-ii). □

**4.7.6 Corollary**

The conditions ($A^*$-i), ($A^*$-iii), ($A^*$-iv) and ($A^*$-v) imply condition ($A^*$-ii). □

# Chapter 5

# Hypoellipticity

We remark that the notion of "hypoellipticity" chosen by P. Kuchment in [Kuc93] Chapter 5 does not coincide with the "usual" definition (e. g. cf. [Hör61] Section 1). We follow P. Kuchment's usage of the notion: "Hypoellipticity" is used to refer to the regularity statement of the forthcoming Theorem 5.1.7, not to a well-defined class of operators.

**5.1.7 Theorem**

Let ($A$-iv) hold. (We remind the reader that condition ($A$-iv) is formulated on page 26 and implies condition ($A^*$-iv), cf. Proposition 4.7.1.)

Furthermore, let $u$ be a solution and assume that $u$ is $z$-quasiperiodic for some $z \in \mathbb{C}\backslash\{0\}$.

Restricting $u$ to $[0,1]$ we have $u \in \mathcal{W}_{1,z}[0,1]$ and $\mathcal{L}_{\mathcal{W}_{1,z}[0,1]}u = 0$.

*Remarks on the proof.*

We extend the proof of [Kuc93] Theorem 5.1.5 to the given situation.

*Proof.*

Proposition 4.5.4 yields $u \in \mathcal{W}_0[0,1]$ and $\widetilde{\mathcal{L}_z} u = 0$.

Let $\delta$, $N$, $(U_j)_{j=1,\dots,N}$, $(t_j)_{j=1,\dots,N}$, $(\phi_j)_{j=1,\dots,N}$, $(\psi_j)_{j=1,\dots,N}$, $(\widetilde{\psi}_j)_{j=1,\dots,N}$ and $\omega(\delta)$ be as in Theorem 3.1.6.

Let $\tilde{\rho} := \rho + \log_{(z)} z$. Then the operators

$$\mathcal{L}_{t_j} := \partial + \mathcal{A}_{t_j} - \tilde{\rho} \in \mathscr{L}(\mathcal{W}_{1,z}[0,1], \mathcal{W}_0[0,1])$$

have a bounded inverse $\mathcal{B}_{t_j} \in \mathscr{L}(\mathcal{W}_0[0,1], \mathcal{W}_{1,z}[0,1])$ for each $j = 1, \dots, N$: Indeed, if we denote by $\mathcal{L}_{t_j,\mathcal{W}_1(\mathbb{T})}$ the operator $\partial + \mathcal{A}_{t_j} - \rho \in \mathscr{L}(\mathcal{W}_1(\mathbb{T}), \mathcal{W}_0[0,1])$ then by Fact 3.1.8 $\mathcal{L}_{t_j} = \mathrm{M}[z_{(z)}^{(\cdot)}](\mathcal{L}_{t_j,\mathcal{W}_1(\mathbb{T})})\mathrm{M}[z_{(z)}^{(-\cdot)}]$ and all three operators on the right hand side are invertible, cf. Theorem 3.1.3 and footnote 9 on page 31.

Next, we will show that $\mathcal{B}_{t_j}$ has an extension to $\mathcal{B}_{t_j}^{\sim} \in \mathscr{L}(\mathcal{W}'_{-1,z}[0,1], \mathcal{W}_0[0,1])$ for each $j = 1, \ldots, N$. The results of Section 4.3 (applied to the (constant) families $(A_{t_j} - \tilde{\rho})_{t\in\mathbb{T}}$ and $(A^*_{t_j} - \tilde{\rho})_{t\in\mathbb{T}})$ yield that $\mathcal{L}_{t_j}$ has an extension $\mathcal{L}_{t_j}^{\sim} \in \mathscr{L}(\mathcal{W}_0[0,1], \mathcal{W}'_{-1,z}[0,1])$ that is given by the dual of the operator $\mathcal{L}'_{t_j} := -\partial + \mathcal{A}'_{t_j} - \tilde{\rho} \in \mathscr{L}(\mathcal{W}'_{1,1/z}[0,1], \mathcal{W}'_0[0,1])$. Analogously as above, if we denote by $\mathcal{L}'_{t_j,\mathcal{W}'_1(\mathbb{T})}$ the operator $-\partial + \mathcal{A}'_{t_j} - \rho \in \mathscr{L}(\mathcal{W}'_1(\mathbb{T}), \mathcal{W}'_0[0,1])$, then $\mathcal{L}'_{t_j} = \mathrm{M}[(1/z)^{(\cdot)}_{(1/z)}](\mathcal{L}_{t_j,\mathcal{W}_1(\mathbb{T})} - \log_{(z)} z - \log_{(1/z)}(1/z))\mathrm{M}[(1/z)^{(-\cdot)}_{(1/z)}]$. Obviously, $-\log_{(1/z)}(1/z)$ is a logarithm of $z$, therefore $c_{\log} := -\log_{(z)} z - \log_{(1/z)}(1/z) \in 2\pi i\mathbb{Z}$. Thus $\rho + i\mathbb{R} \subset \rho(A^*_t + c_{\log})$ and it can be easily shown that $\{ (|\lambda| + 1)(A^*_t + c_{\log} - \lambda)^{-1} : \lambda \in \rho + i\mathbb{R} \} = \{ (|\lambda + c_{\log}| + 1)(A^*_t - \lambda)^{-1} : \lambda \in \rho + i\mathbb{R} \}$ is uniformly $(X^*, X^*)$-$R$-bounded for all $t \in \mathbb{T}$. By Corollary 4.6.2 we conclude that $\mathcal{L}_{t_j,\mathcal{W}_1(\mathbb{T})} - \log_{(z)} z - \log_{(1/z)}(1/z) = -\partial + (\mathcal{A}'_{t_j} + c_{\log}) - \rho$ is invertible and the norm of the inverse can be estimated independently of $t_j$. It follows that $\mathcal{L}'_{t_j}$ has a bounded inverse $\mathcal{B}'_{t_j} \in \mathscr{L}(\mathcal{W}'_0[0,1], \mathcal{W}'_{1,1/z}[0,1])$, whose norm can be estimated by a constant $c_{\mathcal{B}'} > 0$ independently of $t_j$. It is easily checked that $\mathcal{B}_{t_j}^{\sim} := (\mathcal{B}'_{t_j})^* \in \mathscr{L}(\mathcal{W}'_{-1,z}[0,1], \mathcal{W}_0[0,1])$ is an extension of $\mathcal{B}_{t_j}$. We remark that $\|\mathcal{B}_{t_j}^{\sim}\|_{\mathscr{L}(\mathcal{W}'_{-1,z}[0,1],\mathcal{W}_0[0,1])} = \|\mathcal{B}'_{t_j}\|_{\mathscr{L}(\mathcal{W}'_0[0,1],\mathcal{W}'_{1,1/z}[0,1])} \le c_{\mathcal{B}'}$ for all $j = 1, \ldots, N$.

We can again assume that $\delta$ is small enough such that, analogously as in the proof of Theorem 4.6.4, $\mathrm{Id}_{\mathcal{W}_{1,z}[0,1]} + \mathcal{B}_{t_j}(\mathcal{A} - \mathcal{A}_{t_j})\mathrm{M}[\widetilde{\psi_j}] \in \mathscr{L}(\mathcal{W}_{1,z}[0,1])$ is invertible and we set $\mathcal{S}_j := \left(\mathrm{Id}_{\mathcal{W}_{1,z}[0,1]} + \mathcal{B}_{t_j}(\mathcal{A} - \mathcal{A}_{t_j})\mathrm{M}[\widetilde{\psi_j}]\right)^{-1} \in \mathscr{L}(\mathcal{W}_{1,z}[0,1])$.

Also, similarly to the proof of Theorem 4.6.4 (in combination with a duality argument), we obtain $(\mathcal{A}^{\sim} - \mathcal{A}_{t_j}^{\sim})\mathrm{M}[\widetilde{\psi_j}] \in \mathscr{L}(\mathcal{W}_0[0,1], \mathcal{W}'_{-1,z}[0,1])$ and $\|(\mathcal{A}^{\sim} - \mathcal{A}_{t_j}^{\sim})\mathrm{M}[\widetilde{\psi_j}]\|_{\mathscr{L}(\mathcal{W}_0[0,1],\mathcal{W}'_{-1,z}[0,1])} \le \omega(\delta)$ for each $j = 1, \ldots, N$. Therefore $\|\mathcal{B}_{t_j}^{\sim}(\mathcal{A}^{\sim} - \mathcal{A}_{t_j}^{\sim})\mathrm{M}[\widetilde{\psi_j}]\|_{\mathscr{L}(\mathcal{W}_0[0,1])} \le c_{\mathcal{B}'}\omega(\delta) \xrightarrow{\delta\to 0} 0$ and thus we can again assume that $\delta$ is small enough such that $\mathcal{B}_{t_j}^{\sim}(\mathcal{A}^{\sim} - \mathcal{A}_{t_j}^{\sim})\mathrm{M}[\widetilde{\psi_j}]$ is a contraction in $\mathcal{W}_0[0,1]$. Thus then $\mathrm{Id}_{\mathcal{W}_0[0,1]} + \mathcal{B}_{t_j}^{\sim}(\mathcal{A}^{\sim} - \mathcal{A}_{t_j}^{\sim})\mathrm{M}[\widetilde{\psi_j}]$ is invertible and we set $\mathcal{S}_j^{\sim} := \left(\mathrm{Id}_{\mathcal{W}_0[0,1]} + \mathcal{B}_{t_j}^{\sim}(\mathcal{A}^{\sim} - \mathcal{A}_{t_j}^{\sim})\mathrm{M}[\widetilde{\psi_j}]\right)^{-1} \in \mathscr{L}(\mathcal{W}_0[0,1])$.

Then $\mathcal{S}_j^{\sim}$ is an extension of $\mathcal{S}_j$.

We remark that a simple density argument shows that the dual operator $(\mathrm{M}[\psi_j])^* \in \mathscr{L}(\mathcal{W}'_{-1,z}[0,1])$ of $\mathrm{M}[\psi_j] \in \mathscr{L}(\mathcal{W}'_{1,1/z}[0,1])$ is an extension of $\mathrm{M}[\psi_j] \in \mathscr{L}(\mathcal{W}'_0[0,1])$. Thus by the corresponding property of the unextended operators analogous to the proof of Theorem 3.1.6 we obtain by a density argument

$$\Big(\sum_{j=1}^{N} \mathrm{M}[\phi_j]\mathcal{S}_j^{\sim}\mathcal{B}_{t_j}^{\sim}(\mathrm{M}[\psi_j])^*\Big)\mathcal{L}_z^{\sim} =$$

$$\mathrm{Id}_{\mathcal{W}_0[0,1]} + \tilde{\rho}\sum_{j=1}^{N} \phi_j\mathcal{S}_j\mathcal{B}_{t_j}\mathrm{M}[\psi_j] - \sum_{j=1}^{N} \mathrm{M}[\phi_j]\mathcal{S}_j\mathcal{B}_{t_j}\mathrm{M}[\partial\psi_j] \in \mathscr{L}(\mathcal{W}_0[0,1]).$$

Furthermore,

$$\mathcal{R} :=$$

$$\tilde{\rho}\sum_{j=1}^{N} \mathrm{M}[\phi_j]\mathcal{S}_j\mathcal{B}_{t_j}\mathrm{M}[\psi_j] - \sum_{j=1}^{N} \mathrm{M}[\phi_j]\mathcal{S}_j\mathcal{B}_{t_j}\mathrm{M}[\partial\psi_j] \in \mathscr{L}(\mathcal{W}_0[0,1], \mathcal{W}_{1,z}[0,1]).$$

Thus $0 = \Big(\sum_{j=1}^{N} \mathrm{M}[\phi_j]\mathcal{S}_j^{\sim}\mathcal{B}_{t_j}^{\sim}(\mathrm{M}[\psi_j])^*\Big)\mathcal{L}_z^{\sim} u = u + \mathcal{R}u$ and therefore $u = -\mathcal{R}u \in \mathcal{W}_{1,z}[0,1]$.

Then, finally, $\mathcal{L}_{\mathcal{W}_{1,z}[0,1]}u = \mathcal{L}_z^{\sim} u = 0$. □

# Chapter 6

# Basic Properties of Solutions

## 6.1 Functions of Floquet Form

**6.1.1 Definition**
$u \in \mathcal{W}_{0,\mathrm{loc}}(\mathbb{R})$ is called *at most exponentially increasing* if there exist $c, a > 0$ such that $\|u\|_{\mathcal{W}_0[k,k+1]} \leq c \exp(a|k|)$ for all $k \in \mathbb{Z}$.

**6.1.2 Definition**
$u \in \mathcal{W}_{0,\mathrm{loc}}(\mathbb{R})$ is called *of Floquet form* if there are $\lambda \in \mathbb{C}$, $n \in \mathbb{N}_0$ and for each $l = 0, \ldots, n$ $g_l \in L_p(\mathbb{T}, X)$ such that $u = [t \mapsto \exp(\lambda t) \sum_{l=0}^{n} t^l g_l(t)]$ a. e. on $\mathbb{R}$. If $u \neq 0$, then $\exp(\lambda)$ is called the[1] *Floquet exponent* of $u$.

Furthermore, let $z \in \mathbb{C}$.

We denote by $\mathcal{F}form_z$ the union of $0 \in \mathcal{W}_{0,\mathrm{loc}}(\mathbb{R})$ and the set of all non-zero $\mathcal{W}_{0,\mathrm{loc}}(\mathbb{R})$ functions of Floquet form with Floquet exponent $z$.

$u \in \mathcal{F}form_z$ is called *of Bloch form* if there are $\lambda \in \mathbb{C}$ and $g \in L_p(\mathbb{T}, X)$ such that $u = [t \mapsto \exp(\lambda t) g(t)]$ a. e. on $\mathbb{R}$.

We set $\mathcal{B}form_z := \{\, u \in \mathcal{F}form_z : u \text{ is of Bloch form} \,\}$.

The following three facts are direct consequences of the definition or can be easily checked.

**6.1.3 Fact**
$\mathcal{F}form_0 = \{0\}$.

[1] Cf. Proposition 6.1.6.

**6.1.4 Fact**
Every function of Floquet form is at most exponentially increasing.

**6.1.5 Fact**
Let $z \in \mathbb{C}\backslash\{0\}$ and $u \in \mathcal{W}_{0,\mathrm{loc}}(\mathbb{R})$.
Then $u \in \mathcal{B}\mathit{form}_z$ iff $u$ is $z$-quasiperiodic.

**6.1.6 Proposition** (*On Uniqueness of the Floquet Form*)
Let $z \in \mathbb{C}\backslash\{0\}$ and $0 \neq u \in \mathcal{F}\mathit{form}_z$. Let $u = [t \mapsto \exp(\lambda t) \sum_{l=0}^{n} t^l g_l(t)]$ a. e. on $\mathbb{R}$ for some $\lambda \in \mathbb{C}$, $n \in \mathbb{N}_0$ and $g_l \in L_p(\mathbb{T}, X)$ for each $l = 0, \ldots, n$.

Then $\lambda$ is a logarithm of $z$.

In particular, the Floquet exponent of a non-zero function of Floquet form is uniquely determined, i. e. $\mathcal{F}\mathit{form}_{z_1} \cap \mathcal{F}\mathit{form}_{z_2} = \{0\}$ for all $z_1, z_2 \in \mathbb{C}\backslash\{0\}$ with $z_1 \neq z_2$.

Conversely, if $\tilde{\lambda} \in \mathbb{C}$ is a logarithm of $z$ then there are $\tilde{n} \in \mathbb{N}_0$ and $\tilde{g}_l \in L_p(\mathbb{T}, X)$ for each $l = 0, \ldots, \tilde{n}$ such that $u = [t \mapsto \exp(\tilde{\lambda} t) \sum_{l=0}^{\tilde{n}} t^l \tilde{g}_l(t)]$ a. e. on $\mathbb{R}$. Furthermore, if in the above representations $n$ and $\tilde{n}$ are chosen such that $g_n \neq 0$ and $\tilde{g}_{\tilde{n}} \neq 0$ then $\tilde{n} = n$ and $T_{(\lambda,\tilde{\lambda})} g_l = \tilde{g}_l$ for each $l = 0, \ldots, n$ where $T_{(\lambda,\tilde{\lambda})} := [t \mapsto \exp((\lambda - \tilde{\lambda})t)] \in C(\mathbb{R}, \mathbb{C})$ is periodic. In this sense, $n$ is uniquely determined and for each $l = 0, \ldots, \tilde{n}$ $g_l$ is uniquely determined up to a phase shift.

*Proof.*
Let $u = [t \mapsto \exp(\tilde{\lambda} t) \sum_{l=0}^{\tilde{n}} t^l \tilde{g}_l(t)]$ a. e. on $\mathbb{R}$ for some $\tilde{\lambda} \in \mathbb{C}$, $\tilde{n} \in \mathbb{N}_0$ and $\tilde{g}_l \in L_p(\mathbb{T}, X)$ for each $l = 0, \ldots, \tilde{n}$. W. l. o. g. we can assume that $g_n \neq 0$ and $\tilde{g}_{\tilde{n}} \neq 0$. There is a null set $N$ in $\mathbb{R}$ such that (from now on fixed representants of[2]) $t \mapsto \exp(\lambda t) \sum_{l=0}^{n} t^l g_l(t)$ and $t \mapsto \exp(\tilde{\lambda} t) \sum_{l=0}^{\tilde{n}} t^l \tilde{g}_l(t)$ coincide pointwise on $\mathbb{R} \setminus N$. W. l. o. g. we can assume that[3] $N$ is a quasiperiodicity null set for each $g_l$ and $\tilde{g}_l$ for each $l = 0, \ldots, n$ and $l = 0, \ldots, \tilde{n}$, resp.. We set $R := \mathbb{R} \setminus N$. For all $t \in R$ and $k \in \mathbb{Z}$ we set $p_t(k) := \sum_{l=0}^{n} (t+k)^l g_l(t)$ and $\tilde{p}_t(k) := \sum_{l=0}^{\tilde{n}} (t+k)^l \tilde{g}_l(t)$. Thus for every $t \in R$ by periodicity $\left(T_{(\lambda,\tilde{\lambda})} p_t(k)\right)_{k\in\mathbb{Z}} = \left(\exp(-(\lambda - \tilde{\lambda})k)\tilde{p}_t(k)\right)_{k\in\mathbb{Z}}$. There is $\tau \in R$ such that $g_n(\tau) \neq 0$ and by asymptotic analysis we conclude $\operatorname{Re}\lambda = \operatorname{Re}\tilde{\lambda}$ and then

[2] In particular, we fix representants of $g_l$ for each $l = 0, \ldots, n$ and $\tilde{g}_l$ for each $l = 0, \ldots, \tilde{n}$.

[3] Indeed, if $N_l$ and $\tilde{N}_l$ are quasiperiodicity null sets for $g_l$ and $\tilde{g}_l$ for each $l = 0, \ldots, n$ and $l = 0, \ldots, \tilde{n}$, resp., then we can use the null set $\bigcup_{k\in\mathbb{Z}}(k + (N \cup \bigcup_{l=0}^{n} N_l \cup \bigcup_{l=0}^{\tilde{n}} \tilde{N}_l)) \supset N$ instead of $N$.

$\tilde{n} \geq n$. Thus $\left(T_{(\lambda,\tilde{\lambda})}(\tau)\exp(ik\operatorname{Im}(\lambda-\tilde{\lambda}))p_\tau(k)\right)_{k\in\mathbb{Z}} = \left(\tilde{p}_\tau(k)\right)_{k\in\mathbb{Z}}$ and $\tilde{g}_n(\tau) = \lim_{k\to\infty}(\tau+k)^{-n}\tilde{p}_\tau(k) = T_{(\lambda,\tilde{\lambda})}(\tau)\lim_{k\to\infty}\exp(ik\operatorname{Im}(\lambda-\tilde{\lambda}))(\tau+k)^{-n}p_\tau(k) = T_{(\lambda,\tilde{\lambda})}(\tau)\left(\lim_{k\to\infty}\exp(ik\operatorname{Im}(\lambda-\tilde{\lambda}))\right)g_n(\tau)$. $T_{(\lambda,\tilde{\lambda})}(\tau)g_n(\tau) \neq 0$ yields $\operatorname{Im}(\lambda-\tilde{\lambda}) \in 2\pi\mathbb{Z}$, since $\operatorname{Im}(\lambda-\tilde{\lambda}) \notin 2\pi\mathbb{Z}$ would imply that $\left(\exp(ik\operatorname{Im}(\lambda-\tilde{\lambda}))\right)_{k\in\mathbb{Z}}$ has at least two accumulation points. Therefore $T_{(\lambda,\tilde{\lambda})}$ is periodic and for all $t\in\mathbb{R}$ $\left(T_{(\lambda,\tilde{\lambda})}(t)p_t(k)\right)_{k\in\mathbb{Z}} = (\tilde{p}_t(k))_{k\in\mathbb{Z}}$ and thus $\tilde{n} = n$ and $T_{(\lambda,\tilde{\lambda})}g_l = \tilde{g}_l$ (in $L_p(\mathbb{T},X)$) for each $l = 0,\ldots,n$.

Since by definition of $0 \neq u \in \mathcal{F}form_z$ there is a representation where $\tilde{\lambda}$ is a logarithm of $z$ we conclude that $\lambda$ is a logarithm of $z$. In particular, $u \notin \mathcal{F}form_\xi$ for every $\xi \in \mathbb{C}\backslash\{0\}$ with $\xi \neq z$.

Conversely, if $\tilde{\lambda} \in \mathbb{C}$ is a logarithm of $z$ we choose $\lambda \in \mathbb{C}$, $n \in \mathbb{N}_0$ and $g_l \in L_p(\mathbb{T},X)$ for each $l = 0,\ldots,n$ as in Definition 6.1.2. Then again $T_{(\lambda,\tilde{\lambda})}$ is periodic and thus $\tilde{g}_l := T_{(\lambda,\tilde{\lambda})}g_l \in L_p(\mathbb{T},X)$ for each $l = 0,\ldots,n$ and $u = [t \mapsto \exp(\tilde{\lambda}t)\sum_{l=0}^n t^l\tilde{g}_l(t)]$ a. e. on $\mathbb{R}$. □

**6.1.7 Remark**
Obviously, for every $z \in \mathbb{C}$ $\mathcal{F}form_z$ is a linear subspace of $\mathcal{W}_{0,\mathrm{loc}}(\mathbb{R})$.

## 6.2 Solutions – Part 2

**6.2.1 Definition**
Let $z \in \mathbb{C}\backslash\{0\}$.

If $0 \neq u \in \mathcal{F}form_z$ is a solution to (E) then $u$ is called a *Floquet solution.*

If $0 \neq u \in \mathcal{B}form_z$ is a solution to (E) then $u$ is called a *Bloch solution.*

**6.2.2 Definition**
We denote by $\mathcal{F}set$ the set of all $z \in \mathbb{C}\backslash\{0\}$ such that there is a Floquet solution with Floquet exponent $z$. For each $z \in \mathcal{F}set$ we denote by $\mathcal{F}sol_z$ the set of all Floquet solutions with Floquet exponent $z$.

We denote by $\mathcal{B}set$ the set of all $z \in \mathbb{C}\backslash\{0\}$ such that there is a Bloch solution with Floquet exponent $z$. For each $z \in \mathcal{B}set$ we denote by $\mathcal{B}sol_z$ the set of all Bloch solutions with Floquet exponent $z$.

**6.2.3 Remark**
Clearly, $\mathcal{B}set \subset \mathcal{F}set$ and $\mathcal{B}sol_z \subset \mathcal{F}sol_z$ for each $z \in \mathcal{B}set$.

**6.2.4 Remark**
Theorem 5.1.7 in combination with Fact 6.1.5 yields that Bloch solutions are continuous (i. e. they have a continuous representant).

## 6.3 The Test Function Spaces $\Phi_{0,\alpha}$, $\Phi_{1,\alpha}$ $\Phi_0$, $\Phi_1$

We will now define function spaces, that will play the role of test functions on which Floquet solutions will act (similar to distributions) as linear functionals. More precisely, primarily the spaces $\Phi_0'$ and $\Phi_1'$ corresponding to the dual objects (which will be defined in the ) will be used as test spaces (cf. Section 7.1). However, in Theorem 8.1.6 the "predual" versions $\Phi_0$ and $\Phi_1$ will occur and in Theorem 8.1.9 we'll make use of the more general definitions $\Phi_{0,\alpha}'$ and $\Phi_{1,\alpha}'$. Thus for the sake of completeness, we treat the "predual" case in full detail in this section and resume the dual situation in the following section.

**6.3.1 Definition and Proposition** (*The Fréchet Spaces* $\Phi_{0,\alpha}$, $\Phi_{1,\alpha}$ $\Phi_0$, $\Phi_1$)
Let $j \in \{0,1\}$.

For all $a > 0$ and $\phi \in \mathcal{W}_{j,\mathrm{loc}}(\mathbb{R})$ we set
$\gamma_j^{(a)}(\phi) := \sup_{k\in\mathbb{Z}} \|\phi\|_{\mathcal{W}_j[k,k+1]} \exp(a|k|) \in \mathbb{R} \cup \{\infty\}$.

Furthermore, let $\alpha \in (0,\infty]$.

We define $\Phi_{j,\alpha} := \{\, \phi \in \mathcal{W}_{j,\mathrm{loc}}(\mathbb{R}) : \gamma_j^{(a)}(\phi) < \infty \text{ for all } a \in (0,\alpha) \,\}$.

We set $\Phi_j := \Phi_{j,\infty}$.

For every $a > 0$ $\gamma_j^{(a)}$ is a seminorm on $\Phi_{j,\alpha}$ and we endow $\Phi_{j,\alpha}$ with the topology generated by the family $\{\gamma_j^{(a)}\}_{a>0}$.

Then $\Phi_{j,\alpha}$ is a Fréchet space.

*Proof.*
The vector space structure of $\Phi_{j,\alpha}$ is immediate and obviously $\gamma_j^{(a)}$ is a norm for each $a > 0$. In particular, $\Phi_{j,\alpha}$ is Hausdorff. Furthermore, if $b \geq a > 0$ then $\gamma_j^{(b)} \geq \gamma_j^{(a)}$. Therefore, the topology on $\Phi_j$ is already generated by the family $\{\gamma_j^{(n)}\}_{n\in\mathbb{N}}$, and if $\alpha < \infty$ the topology on $\Phi_{j,\alpha}$ is already generated by the countable family $\{\gamma_j^{(\alpha-1/n)}\}_{n\in\mathbb{N}}$. In any case, the topology on $\Phi_{j,\alpha}$ is generated by a countable family.

It remains to show completeness. Let $(\phi_n)_{n\in\mathbb{N}}$ be a Cauchy sequence in $\Phi_{j,\alpha}$. An easy calculations shows that $\Phi_{j,\alpha} \hookrightarrow \mathcal{W}_{j,\text{loc}}(\mathbb{R})$ and we conclude that $(\phi_n)_{n\in\mathbb{N}}$ is a Cauchy sequence in $\mathcal{W}_{j,\text{loc}}(\mathbb{R})$ and therefore converges in $\mathcal{W}_{j,\text{loc}}(\mathbb{R})$ to, say, $\phi_0$.

We will now show that $\gamma_j^{(a)}(\phi_n - \phi_0) \overset{n\to\infty}{\longrightarrow} 0$ for all $a \in (0,\alpha)$. Let $a \in (0,\alpha)$ and $\epsilon > 0$. There is $N > 0$ such that $\gamma_j^{(a)}(\phi_n - \phi_m) < \epsilon/2$ if $n, m > N$. Furthermore, for each $k \in \mathbb{Z}$ there is $N_k \geq N$ such that $\|\phi_{N_k} - \phi_0\|_{\mathcal{W}_j[k,k+1]} \exp(a|k|) < \epsilon/2$. We obtain for all $n > N$ and all $k \in \mathbb{Z}$ $\|\phi_n - \phi_0\|_{\mathcal{W}_j[k,k+1]} \exp(a|k|) \leq \gamma_j^{(a)}(\phi_n - \phi_{N_k}) + \|\phi_{N_k} - \phi_0\|_{\mathcal{W}_j[k,k+1]} \exp(a|k|) < \epsilon$. In particular, $\gamma_j^{(a)}(\phi_0) < \infty$. Thus $\phi_0 \in \Phi_{j,\alpha}$ and $\phi_n \underset{\Phi_{j,\alpha}}{\overset{n\to\infty}{\longrightarrow}} \phi_0$. Hence $\Phi_{j,\alpha}$ is complete. □

The following remark indicates that these functions are suitable as coefficients of Laurent series on $\mathbb{C}\backslash\{0\}$ about 0, cf. Fact 1.5.15. We will make us of that observation (more precisely: of the analogous statement in the dual situation, cf. Remark 6.4.2) in Construction 7.6.9.

### 6.3.2 Remark

Let $j \in \{0,1\}$ and $\alpha \in (0,\infty]$.
For all $\phi \in \Phi_{j,\alpha}$ and $k \in \mathbb{Z}$, clearly $\phi_k := (\phi(\cdot - k))_{|[0,1]} \in \mathcal{W}_j[0,1]$.
A direct calculation shows $\phi \in \Phi_{j,\alpha}$ iff $\phi \in \mathcal{W}_{j,\text{loc}}(\mathbb{R})$ and[4]

$\limsup_{k\to\infty} \left(\|\phi_k\|_{\mathcal{W}_j[0,1]}\right)^{1/k} \leq \exp(-\alpha)$ and

$\limsup_{k\to\infty} \left(\|\phi_{-k}\|_{\mathcal{W}_j[0,1]}\right)^{1/k} \leq \exp(-\alpha)$.

### 6.3.3 Remark

Let $j \in \{0,1\}$ and $\alpha, \beta \in (0,\infty]$ with $\alpha \leq \beta$.
Then obviously, $\Phi_{j,\beta} \hookrightarrow \Phi_{j,\alpha}$.

### 6.3.4 Remark

Obviously, $\Phi_1 \hookrightarrow \Phi_0$.

For each $j = 0, 1$ the rapid decay of functions in $\Phi_j$ directly yields that $\mathcal{W}_{j,\text{loc}}(\mathbb{R})$-functions with compact support are dense in $\Phi_j$ and then with a mollifying argument (cf. [Ama95] Section III.4.2) we obtain:

### 6.3.5 Fact

$C_c^\infty(\mathbb{R}, D)$ is dense in $\Phi_1$.

$C_c^\infty(\mathbb{R}, X)$ is dense in $\Phi_0$.

---

[4] Here, we use the convention $\exp(-\infty) := 0$, of course.

## 6.4 The Test Function Spaces $\Phi'_{0,\alpha}$, $\Phi'_{1,\alpha}$ $\Phi'_0$, $\Phi'_1$

As explained in the previous section, we will state the corresponding definitions and results for the dual situation.

**6.4.1 Definition and Proposition** (*The Fréchet Spaces* $\Phi'_{0,\alpha}$, $\Phi'_{1,\alpha}$ $\Phi'_0$, $\Phi'_1$)
Let $j \in \{0,1\}$.
By abuse of notation, for all $a > 0$ and $\phi' \in \mathcal{W}'_{j,\mathrm{loc}}(\mathbb{R})$ we set
$\gamma_j^{(a)}(\phi') := \sup_{k\in\mathbb{Z}} \|\phi'\|_{\mathcal{W}'_j[k,k+1]} \exp(a|k|) \in \mathbb{R} \cup \{\infty\}$.
Furthermore, let $\alpha \in (0,\infty]$.
We define $\Phi'_{j,\alpha} := \{\, \phi' \in \mathcal{W}'_{j,\mathrm{loc}}(\mathbb{R}) : \gamma_j^{(a)}(\phi') < \infty \text{ for all } a \in (0,\alpha) \,\}$.
We set $\Phi'_j := \Phi'_{j,\infty}$.
For every $a > 0$ $\gamma_j^{(a)}$ is a seminorm on $\Phi'_{j,\alpha}$ and we endow $\Phi'_{j,\alpha}$ with the topology generated by the family $\{\gamma_j^{(a)}\}_{a>0}$.
Then $\Phi'_{j,\alpha}$ is a Fréchet space.

**6.4.2 Remark**
Let $j \in \{0,1\}$ and $\alpha \in (0,\infty]$.
Then $\phi' \in \Phi'_{j,\alpha}$ iff $\phi' \in \mathcal{W}'_{j,\mathrm{loc}}(\mathbb{R})$ and[5]
$\limsup_{k\to\infty} \left(\|\phi'_k\|_{\mathcal{W}'_j[0,1]}\right)^{1/k} \le \exp(-\alpha)$ and
$\limsup_{k\to\infty} \left(\|\phi'_{-k}\|_{\mathcal{W}'_j[0,1]}\right)^{1/k} \le \exp(-\alpha)$,
where again $\phi'_k := (\phi'(\cdot - k))_{|[0,1]} \in \mathcal{W}'_j[0,1]$ for all $\phi' \in \Phi'_{j,\alpha}$ and $k \in \mathbb{Z}$.

**6.4.3 Remark**
Let $j \in \{0,1\}$ and $\alpha, \beta \in (0,\infty]$ with $\alpha \le \beta$.
Then $\Phi'_{j,\beta} \hookrightarrow \Phi'_{j,\alpha}$.

**6.4.4 Remark**
$\Phi'_1 \hookrightarrow \Phi'_0$.

**6.4.5 Fact**
$C_c^\infty(\mathbb{R}, D')$ is dense in $\Phi'_1$.
$C_c^\infty(\mathbb{R}, X^*)$ is dense in $\Phi'_0$.

[5]Here, we use again the convention $\exp(-\infty) := 0$.

## 6.5 The Operators $\mathcal{L}$ and $\mathcal{L}'$ on the Real Line – Part 2

### 6.5.1 Definition and Proposition

Let $\alpha \in (0, \infty]$.

We denote by[6] $\mathcal{L}_{\Phi_{1,\alpha}}$ the restriction of $\mathcal{L}_{\mathcal{W}_{1,\mathrm{loc}}(\mathbb{R})}$ to $\Phi_{1,\alpha}$ and by $\mathcal{L}'_{\Phi'_{1,\alpha}}$ the restriction of $\mathcal{L}'_{\mathcal{W}'_{1,\mathrm{loc}}(\mathbb{R})}$ to $\Phi'_{1,\alpha}$.
Then $\mathcal{L}_{\Phi_{1,\alpha}} \in \mathscr{L}(\Phi_{1,\alpha}, \Phi_{0,\alpha})$ and $\mathcal{L}'_{\Phi'_{1,\alpha}} \in \mathscr{L}(\Phi'_{1,\alpha}, \Phi'_{0,\alpha})$.

*Proof.*
We will prove $\mathcal{L}_{\Phi_{1,\alpha}} \in \mathscr{L}(\Phi_{1,\alpha}, \Phi_{0,\alpha})$. The second statement follows analogously. For all $k \in \mathbb{Z}$ obviously

$$\|\mathcal{L}_{\mathcal{W}_1[k,k+1]}\|_{\mathscr{L}(\mathcal{W}_1[k,k+1],\mathcal{W}_0[k,k+1])} = \|\mathcal{L}_{\mathcal{W}_1[0,1]}\|_{\mathscr{L}(\mathcal{W}_1[0,1],\mathcal{W}_0[0,1])},$$

where of course $\mathcal{L}_{\mathcal{W}_1[k,k+1]} := -\partial + \mathcal{A} \in \mathscr{L}(\mathcal{W}_1[k, k+1], \mathcal{W}_0[k, k+1])$. Therefore $\gamma_0^{(a)}(\mathcal{L}_{\Phi_{1,\alpha}}\phi) \le \|\mathcal{L}_{\mathcal{W}_1[0,1]}\|_{\mathscr{L}(\mathcal{W}_1[0,1],\mathcal{W}_0[0,1])} \cdot \gamma_1^{(a)}(\phi)$ for all $a \in (0, \alpha)$ and all $\phi \in \Phi_{1,\alpha}$. This directly yields the assertion. □

### 6.5.2 Definition and Proposition

Let $u \in \mathcal{W}_{0,\mathrm{loc}}(\mathbb{R})$ be at most exponentially increasing with corresponding constants $c, a > 0$, i. e. $\|u\|_{\mathcal{W}_0[k,k+1]} \le c\exp(a|k|)$ for all $k \in \mathbb{Z}$.
Let $\alpha \in (a, \infty]$.
For each $\phi' \in \Phi'_{0,\alpha}$ we set $(\mathrm{F}u)(\phi') := \int_{\mathbb{R}} \langle \phi'(t), u(t)\rangle_X \, dt \in \mathbb{C}$.
Then $\mathrm{F}u \in (\Phi'_{0,\alpha})^*$ and the map $\mathrm{F} : \mathcal{W}_{0,\mathrm{loc}}(\mathbb{R}) \to (\Phi'_{0,\alpha})^*$ is injective. In particular, $\mathrm{F}u \in (\Phi'_0)^*$ and the map $\mathrm{F} : \mathcal{W}_{0,\mathrm{loc}}(\mathbb{R}) \to (\Phi'_0)^*$ is injective.

*Proof.*
First we remark that for each $\phi' \in \Phi'_{0,\alpha}$ by [Edw65] Theorem 8.20.5 $[t \mapsto \phi(t)u(t)] \in L_{1,\mathrm{loc}}(\mathbb{R}, \mathbb{C})$. Let $\tilde{a} \in (a, \alpha)$. Since clearly $\mathrm{F}u$ is linear $\mathrm{F}u \in (\Phi'_0)^*$ follows from

$$|(\mathrm{F}u)(\phi)| \le \int_{\mathbb{R}} |\phi(t)u(t)| \, dt = \sum_{k=-\infty}^{\infty} \int_k^{k+1} |\phi(t)u(t)| \, dt \le$$

$$\sum_{k=-\infty}^{\infty} \|\phi\|_{\mathcal{W}'_0[k,k+1]} \|u\|_{\mathcal{W}_0[k,k+1]} \le \sum_{k=-\infty}^{\infty} \|\phi\|_{\mathcal{W}'_0[k,k+1]} c\exp(a|k|) \le$$

$$c \cdot \big(\sup_{k\in\mathbb{Z}} \|\phi'\|_{\mathcal{W}'_0[k,k+1]} \exp(\tilde{a}|k|)\big)\Big(\sum_{k=-\infty}^{\infty} \exp(-\tilde{a}|k|)\exp(a|k|)\Big) \le$$

[6] Of course, this is meant to include the cases $\mathcal{L}_{\Phi_1} := \mathcal{L}_{\Phi_{1,\infty}}$ and $\mathcal{L}'_{\Phi'_1} := \mathcal{L}'_{\Phi'_{1,\infty}}$.

$c \cdot \frac{\exp(\tilde{a}-a)+1}{\exp(\tilde{a}-a)-1} \cdot \gamma_0^{(\tilde{a})}(\phi')$

for each $\phi' \in \Phi'_{0,\alpha}$.

If $u_1, u_2 \in \mathcal{W}_{0,\mathrm{loc}}(\mathbb{R})$ with $\mathrm{F}u_1 = \mathrm{F}u_2$ then $\int_K \langle \phi'(t), (u_1 - u_2)(t) \rangle_X \, dt = 0$ for all $K \subset\subset \mathbb{R}$ and $\phi' \in C_c^\infty(\mathbb{R}, X^*)$ with $\operatorname{supp} \phi' \subset K$.This yields $u_1 = u_2$ a. e. on $K$ and hence on $\mathbb{R}$. □

### 6.5.3 Proposition

Let $u \in \mathcal{W}_{0,\mathrm{loc}}(\mathbb{R})$ be at most exponentially increasing.
Then $u$ is a solution to (E) iff $\mathrm{F}u \in \operatorname{Coker} \mathcal{L}'_{\Phi'_1}$.

*Proof.*
First, we note that by definition $\mathcal{L}'_{\mathcal{W}'_{1,\mathrm{loc}}(\mathbb{R})}\phi' = \mathcal{L}'_{\Phi'_1}\phi'$ for all $\phi' \in C_c^\infty(\mathbb{R}, D')$ and $\langle \mathcal{L}'_{\mathcal{W}'_{1,\mathrm{loc}}(\mathbb{R})}\phi', u \rangle = (\mathrm{F}u)(\mathcal{L}'_{\Phi'_1}\phi')$. Thus if $\mathrm{F}u \in \operatorname{Coker} \mathcal{L}'_{\Phi'_1}$ (and hence $(\mathrm{F}u)(\mathcal{L}'_{\Phi'_1}\phi') = 0$ for all $\phi' \in C_c^\infty(\mathbb{R}, D')$) this directly yields that $u$ is a solution. Conversely, if $u$ is a solution then, since by Fact 6.4.5 $C_c^\infty(\mathbb{R}, D')$ is dense in $\Phi'_1$, continuity of $\mathrm{F}u$ and $\mathcal{L}'_{\Phi'_1}$ yields $(\mathrm{F}u)(\mathcal{L}'_{\Phi'_1}\phi') = 0$ for all $\phi' \in \Phi'_1$ and therefore $\mathrm{F}u \in \operatorname{Coker} \mathcal{L}'_{\Phi'_1}$. □

# Chapter 7

# Transformation of the Problem

## 7.1 Motivation

We first remark that this motivation is intended to give an overview of the constructions and statements in this chapter. We will clarify the rigorous mathematical meaning in the following sections. In particular, we will give definitions of the mentioned objects. Furthermore, we refer to the appendix for a precise explanation of the structure of analytic bundles and their sections.

We recall that we want to analyze solutions to the equation $\mathcal{L}u = 0$. In order to allow solutions in $\mathcal{W}_{0,\mathrm{loc}}(\mathbb{R})$, in Section 4.5 we have introduced a weak formulation, where we used $C_c^\infty(\mathbb{R}, X^*)$ as a test function space and the "dual" operator $\mathcal{L}'$ acting on that test function space. (We have seen in Proposition 4.5.3 that this weak formulation indeed leads to $\mathcal{L}u = 0$ if $u$ is regular enough.)

However, the test function space $C_c(\mathbb{R}, X^*)$ is not suitable for our needs. If we restrict ourselves to solutions that are at most exponentially increasing then Proposition 6.5.2 allows the bigger test function space $\Phi_1'$. (Indeed, when using $C_c^\infty(\mathbb{R}, X^*)$ as a test function space, functions in the image under $\mathcal{L}'$ still have compact support and therefore allow integration against a $\mathcal{W}_{0,\mathrm{loc}}(\mathbb{R})$-function. With the test function space $\Phi_1'$ functions in the image under $\mathcal{L}'$ decay fast enough (they are in $\Phi_0'$) to allow the integration against an at most exponentially increasing $\mathcal{W}_{0,\mathrm{loc}}(\mathbb{R})$-function.)

After switching to the "right" test function space, "the place to look for solutions" is then the dual space of $\Phi_0'$ and in particular, solutions can now be described as the cokernel of $\mathcal{L}'_{\Phi'_1}$, cf. Proposition 6.5.3.

However, the main benefit of using $\Phi'_1$ is that the problem can be translated into the analysis of the spectrum and cospectrum of certain homomorphism of sections of bundles.

Indeed, we will introduce an isomorphism, namely the Floquet transform $\mathcal{U}$, to carry both the structure of $\Phi'_1$ and $\Phi'_0$ to section spaces $\Gamma(\mathbb{C}\backslash\{0\}, \langle\mathfrak{C}'_1\rangle)$ and $\Gamma(\mathbb{C}\backslash\{0\}, \langle\mathfrak{B}'_0\rangle)$, resp., where $\langle\mathfrak{C}'_1\rangle$ and $\langle\mathfrak{B}'_0\rangle$ will be suitable bundles.

This leads to the map $\Gamma(\mathbb{C}\backslash\{0\}, \langle\mathfrak{C}'_1\rangle) \xrightarrow{\mathcal{U}^{-1}} \Phi'_1 \xrightarrow{\mathcal{L}'} \Phi'_0 \xrightarrow{\mathcal{U}} \Gamma(\mathbb{C}\backslash\{0\}, \langle\mathfrak{B}'_0\rangle)$ between those section spaces. This map can be identified with the induced homomorphism of a bundle homomorphism $\mathfrak{L}'$ between $\langle\mathfrak{C}'_1\rangle$ and $\langle\mathfrak{B}'_0\rangle$, cf. Corollary 7.6.12. (This bundle homomorphism will be constructed explicitly and could be described as the fiberwise action of $\mathcal{L}'$ and it will be a Fredholm homomorphism.)

We will see that solutions of Floquet (and Bloch) form will play an important role in describing all at most exponentially increasing solutions. As mentioned, we will understand solutions as functionals on $\Phi'_0$. Then the Floquet transform (or more precisely: $(\mathcal{U}^*)^{-1}$) will translate those functionals to functionals on $\Gamma(\mathbb{C}\backslash\{0\}, \langle\mathfrak{B}'_0\rangle)$. By Proposition 7.7.6 we will be able to describe the image of Floquet solutions.

The following diagram illustrates the situation.

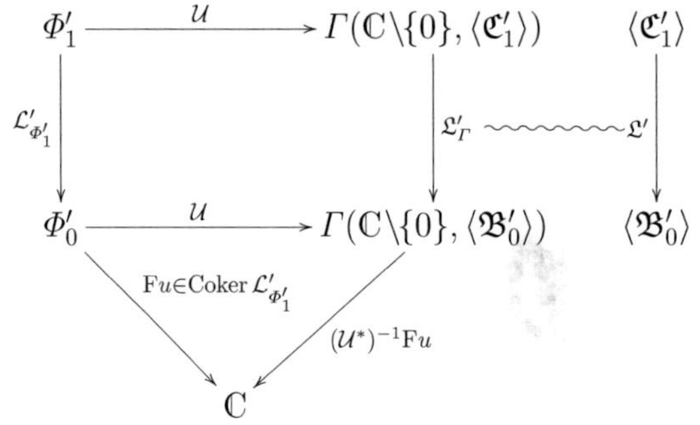


In this situation, Proposition 7.8.1 will yield a correspondence of the set of Floquet exponents to the cospectrum of the bundle homomorphism described above. In particular, we will obtain that the set of Floquet exponents is an analytic set.

Furthermore, thanks to the Fredholm property of the bundle homomorphism a superposition principle for functionals on $\Gamma(\mathbb{C}\backslash\{0\}, \langle\mathfrak{B}_0'\rangle)$ is known, cf. [Kuc93] Theorem 1.7.1. After translating back to the original problem, this will yield the superposition principle for at most exponentially increasing solutions, the central result of this thesis.

## 7.2 The Bundles $\langle\mathfrak{B}_0\rangle$ and $\langle\mathfrak{B}_0'\rangle$

In this section, by abuse of notation we will use the symbol $\nu$ for a projection function $\nu : M \times N \mapsto M$ defined by $\nu(m,n) := m$, where $m \in M$ and $n \in N$, for varying sets $M$ and $N$.

**7.2.1 Construction** (*The Bundles* $\langle\mathfrak{B}_0\rangle$ *and* $\langle\mathfrak{B}_0'\rangle$)
Let $\mathfrak{B}_0 := \mathbb{C}\backslash\{0\} \times \mathcal{W}_0[0,1]$ and $\mathfrak{B}_0' := \mathbb{C}\backslash\{0\} \times \mathcal{W}_0'[0,1]$. Furthermore, let $\mathfrak{b}_0 := \mathrm{Id}_{\mathfrak{B}_0}$ and $\mathfrak{b}_0' := \mathrm{Id}_{\mathfrak{B}_0'}$. We denote by $\langle\mathfrak{B}_0\rangle := \langle\mathfrak{B}_0 \overset{\nu}{\succ} \mathbb{C}\backslash\{0\}\rangle$ and $\langle\mathfrak{B}_0'\rangle := \langle\mathfrak{B}_0' \overset{\nu}{\succ} \mathbb{C}\backslash\{0\}\rangle$, resp., the trivial bundle constructed in Example A.1.8 (with trivialization $\{\mathrm{Id} : \mathfrak{B}_0 \to \mathbb{C}\backslash\{0\} \times \mathcal{W}_0[0,1]\} = \{\mathfrak{b}_0\}$ and with $\Omega = \mathbb{C}\backslash\{0\}$, $\mathcal{E} = \mathfrak{B}_0$ and $[\nu : \mathbb{C}\backslash\{0\} \times \mathcal{W}_0[0,1] \to \mathbb{C}\backslash\{0\}] = [p : \mathcal{E} \to \Omega]$ and analogously for $\langle\mathfrak{B}_0' \overset{\nu}{\succ} \mathbb{C}\backslash\{0\}\rangle$). △

### Duality between $\langle\mathfrak{B}_0'\rangle$ and $\langle\mathfrak{B}_0\rangle$

**7.2.2 Definition and Remark**
Let $z \in \mathbb{C}\backslash\{0\}$. For all $f' \in (\mathfrak{B}_0')_z$ and $f \in (\mathfrak{B}_0)_z$ we set $\langle f', f\rangle_z := \langle(\mathfrak{b}_0')_z f', (\mathfrak{b}_0)_z f\rangle_{\mathcal{W}}$. In view of Fact 4.2.3, obviously $\big((\mathfrak{B}_0)_z\big)^*$ can be identified with $(\mathfrak{B}_0')_z$ and $\big((\mathfrak{B}_0')_z\big)^*$ can be identified with $(\mathfrak{B}_0)_z$ by the isomorphisms $(\mathfrak{B}_0')_z \ni f' \mapsto \langle f', \cdot\rangle_z \in \big((\mathfrak{B}_0)_z\big)^*$ and $(\mathfrak{B}_0)_z \ni f \mapsto \langle \cdot, f\rangle_z \in \big((\mathfrak{B}_0')_z\big)^*$.

**7.2.3 Definition and Proposition**
Let $\emptyset \neq \Omega \overset{\circ}{\subset} \mathbb{C}\backslash\{0\}$, $s' \in \Gamma(\Omega, \langle\mathfrak{B}_0'\rangle)$ and $s \in \Gamma(\Omega, \langle\mathfrak{B}_0\rangle)$.
We set $\langle s', s\rangle_\Omega := [\Omega \ni z \mapsto \langle s'(z), s(z)\rangle_z \in \mathbb{C}]$.
Then $\langle\cdot,\cdot\rangle_\Omega \in \mathscr{L}(\Gamma(\Omega, \langle\mathfrak{B}_0'\rangle); \Gamma(\Omega, \langle\mathfrak{B}_0\rangle), A(\Omega, \mathbb{C}))$.

*Proof.*
This is a direct consequence of Proposition A.2.8 and Fact 1.5.9. □

**7.2.4 Remark**
In view of Remark A.2.9 in the situation of Definition 7.2.3 $\langle\cdot,\cdot\rangle_O$ is well-defined for every $\emptyset \neq O \overset{\circ}{\subset} \Omega$ and coincides with restriction of $\langle\cdot,\cdot\rangle_\Omega$ to sections on $O$.

## 7.3 The Bundles $\langle\mathfrak{B}_1\rangle$ and $\langle\mathfrak{C}_1\rangle$

**7.3.1 Construction** (*The Bundle* $\langle\mathfrak{B}_1\rangle$)
Let $\mathfrak{B}_1 := \mathbb{C}\backslash\{0\} \times \mathcal{W}_1[0,1]$ and $\mathfrak{b}_1 := \mathrm{Id}_{\mathfrak{B}_1}$. Let $\langle\mathfrak{B}_1\rangle := \langle\mathfrak{B}_1 \overset{\nu}{\succ} \mathbb{C}\backslash\{0\}\rangle$ be the trivial bundle constructed in Example A.1.8 (with trivialization $\{\mathrm{Id} : \mathfrak{B}_1 \to \mathbb{C}\backslash\{0\} \times \mathcal{W}_1[0,1]\} = \{\mathfrak{b}_1\}$ and with $\Omega = \mathbb{C}\backslash\{0\}$, $\mathcal{E} = \mathfrak{B}_1$ and $[\nu : \mathbb{C}\backslash\{0\} \times \mathcal{W}_1[0,1] \to \mathbb{C}\backslash\{0\}] = [p : \mathcal{E} \to \Omega]$).
Furthermore, for all $z \in \mathbb{C}\backslash\{0\}$ and $(\xi, f) \in (\mathfrak{B}_1)_{|\mathbb{B}_z}$ $\mathfrak{b}_1{}^{(z)}(\xi, f) := (\xi, \mathrm{M}[\xi_{(z)}^{(-\cdot)}]f)$ defines a homeomorphism $\mathfrak{b}_1{}^{(z)} : (\mathfrak{B}_1)_{|\mathbb{B}_z} \to \mathbb{B}_z \times \mathcal{W}_1[0,1]$. $\{\mathfrak{b}_1{}^{(z)}\}_{z\in\mathbb{C}\backslash\{0\}}$ is a trivialization that is equivalent to $\{\mathfrak{b}_1\}$.

*Proof.*
It can be easily checked that $\mathfrak{b}_1{}^{(z)} : (\mathfrak{B}_1)_{|\mathbb{B}_z} \to \mathbb{B}_z \times \mathcal{W}_1[0,1]$ is a homeomorphism and obviously, condition (a) of Definition A.1.1 holds. Furthermore, for each $\xi \in \mathbb{B}_z$, the induced map $(\mathfrak{b}_1{}^{(z)})_\xi : (\mathfrak{B}_1)_\xi \to \mathcal{W}_1[0,1]$ coincides with the multiplication operator $\mathrm{M}[\xi_{(z)}^{(-\cdot)}] \circ \cong\, \in \mathscr{L}((\mathfrak{B}_1)_\xi, \mathcal{W}_1[0,1])$, where $\cong$ denotes the natural identification of $(\mathfrak{B}_1)_\xi$ with $\mathcal{W}_1[0,1]$, i.e. the isometric isomorphism $(\mathfrak{b}_1)_\xi \in \mathscr{L}((\mathfrak{B}_1)_\xi, \mathcal{W}_1[0,1])$. Thus Remark A.1.4 yields condition (b) of Definition A.1.1. Finally, if $z, \tilde{z} \in \mathbb{C}\backslash\{0\}$ with $\mathbb{B}_z \cap \mathbb{B}_{\tilde{z}} \neq$ then $\mathbb{B}_z \cap \mathbb{B}_{\tilde{z}}$ is a domain and [RS02] Section 5.4.1 yields that there is $k \in \mathbb{Z}$ such that for every $\xi \in \mathbb{B}_z \cap \mathbb{B}_{\tilde{z}}$ $((\mathfrak{b}_1{}^{(z)})_\xi)^{-1}(\mathfrak{b}_1{}^{(\tilde{z})})_\xi$ coincides with the multiplication operator $\mathrm{M}[\exp(2\pi i k(\cdot))] \in \mathscr{L}(\mathcal{W}_1[0,1])$ and thus the corresponding transition function is constant. This yields condition (c) of Definition A.1.1. Thus $\{\mathfrak{b}_1{}^{(z)}\}_{z\in\mathbb{C}\backslash\{0\}}$ is a trivialization and we now show that it is equivalent to $\{\mathfrak{b}_1\}$: Let $z \in \mathbb{C}\backslash\{0\}$. By Remark A.1.2 it suffices to show that the map $\Psi_\xi : \mathcal{W}_1[0,1] \overset{((\mathfrak{b}_1{}^{(z)})_\xi)^{-1}}{\longrightarrow} (\mathfrak{B}_1)_\xi \overset{\mathrm{Id}_\xi}{\longrightarrow} \mathcal{W}_1[0,1]$ is analytically depending on $\xi \in \mathbb{B}_z$. Since for every $\xi \in \mathbb{B}_z$ $\Psi_\xi$ coincides with the multiplication operator $\mathrm{M}[\xi_{(z)}^{(-\cdot)}] \in \mathscr{L}(\mathcal{W}_1[0,1])$, Fact 1.9.1, Fact 1.8.3 and Fact 1.5.10 yield

$[\xi \mapsto \Psi_\xi] \in A(\mathbb{B}_z, \mathscr{L}(\mathcal{W}_1[0,1]))$. □

**7.3.2 Construction** (*The Bundle* $\langle\mathfrak{C}_1\rangle$)
We endow $\mathfrak{C}_1 := \bigcup_{z\in\mathbb{C}\setminus\{0\}}\{z\} \times \mathcal{W}_{1,z}[0,1] \subset \mathfrak{B}_1$ with the induced topology.

By (further) abuse of notation we denote the restriction of $\nu : \mathfrak{B}_1 \to \mathbb{C}\setminus\{0\}$ to $\mathfrak{C}_1$ also by $\nu$.

Furthermore, for all $z \in \mathbb{C}\setminus\{0\}$ we understand $\{z\} \times \mathcal{W}_{1,z}[0,1] = (\mathfrak{C}_1)_z = (\nu_{|\mathfrak{C}_1})^{-1}(z)$ as a (closed) linear subspace of $(\mathfrak{B}_1)_z$. (Thus $(\mathfrak{C}_1)_z$ has a Banach space structure and it coincides with the natural identification with $\mathcal{W}_{1,z}[0,1]$.)

Then $\nu : \mathfrak{C}_1 \to \mathbb{C}\setminus\{0\}$ is a bundle projection.

For each $z \in \mathbb{C}\setminus\{0\}$ let $\mathfrak{c}_1{}^{(z)}$ be the restriction of $\mathfrak{b}_1{}^{(z)}$ to $(\mathfrak{C}_1)_{|\mathbb{B}_z}$. Then $\mathfrak{c}_1{}^{(z)} : (\mathfrak{C}_1)_{|\mathbb{B}_z} \to \mathbb{B}_z \times \mathcal{W}_1(\mathbb{T})$ is a homeomorphism. $\{\mathfrak{c}_1{}^{(z)}\}_{z\in\mathbb{C}\setminus\{0\}}$ is a trivialization.

For all $z \in \mathbb{C}\setminus\{0\}$ and $\xi \in \mathbb{B}_z$ the induced map $(\mathfrak{c}_1{}^{(z)})_\xi : (\mathfrak{C}_1)_\xi \to \mathcal{W}_1(\mathbb{T})$ coincides with the multiplication operator $\mathrm{M}[\xi_{(z)}^{(-\cdot)}] \circ \cong \in \mathscr{L}((\mathfrak{C}_1)_\xi, \mathcal{W}_1(\mathbb{T}))$, where $\cong$ denotes the natural identification of $(\mathfrak{C}_1)_\xi$ with $\mathcal{W}_{1,\xi}[0,1]$, i.e. the isometric isomorphism $((\mathfrak{b}_1)_\xi)_{|(\mathfrak{C}_1)_\xi} \in \mathscr{L}((\mathfrak{C}_1)_\xi, \mathcal{W}_{1,\xi}[0,1])$.

The equivalence class of $\{\mathfrak{c}_1{}^{(z)}\}_{z\in\mathbb{C}\setminus\{0\}}$ is denoted by $\langle\mathfrak{C}_1\rangle$.

*Proof.*
Obviously, for every $z \in \mathbb{C}\setminus\{0\}$, the topology of $(\mathfrak{C}_1)_z$ coincides with the topology induced by the Banach space structure of $(\mathfrak{C}_1)_z$ and therefore $\nu : \mathfrak{C}_1 \to \mathbb{C}\setminus\{0\}$ is a bundle projection. Next, we note that for each $z \in \mathbb{C}\setminus\{0\}$ $\mathfrak{c}_1{}^{(z)}$ maps indeed into $\mathbb{B}_z \times \mathcal{W}_1(\mathbb{T})$ since $(\mathrm{M}[\xi_{(z)}^{(-\cdot)}]f)(0) = f(0) = \xi^{-1}f(1) = (\mathrm{M}[\xi_{(z)}^{(-\cdot)}]f)(1)$ for all $(\xi, f) \in (\mathfrak{C}_1)_{|\mathbb{B}_z}$ and thus by Fact 1.8.3 $\mathrm{M}[\xi_{(z)}^{(-\cdot)}]f \in \mathcal{W}_1(\mathbb{T})$. We conclude that $\mathfrak{c}_1{}^{(z)} : (\mathfrak{C}_1)_{|\mathbb{B}_z} \to \mathbb{B}_z \times \mathcal{W}_1(\mathbb{T})$ is a homeomorphism and that the induced map $(\mathfrak{c}_1{}^{(z)})_\xi : (\mathfrak{C}_1)_\xi \to \mathcal{W}_1(\mathbb{T})$ coincides with the multiplication operator $\mathrm{M}[\xi_{(z)}^{(-\cdot)}] \circ \cong \in \mathscr{L}((\mathfrak{C}_1)_\xi, \mathcal{W}_1(\mathbb{T}))$. Again, condition (a) of Definition A.1.1 holds and Remark A.1.4 yields condition (b) of Definition A.1.1. Analogously as in Construction 7.4.1, if $z, \tilde{z} \in \mathbb{C}\setminus\{0\}$ with $\mathbb{B}_z \cap \mathbb{B}_{\tilde{z}} \neq$ then there is $k \in \mathbb{Z}$ such that for every $\xi \in \mathbb{B}_z \cap \mathbb{B}_{\tilde{z}}$ $((\mathfrak{c}_1{}^{(z)})_\xi)^{-1}(\mathfrak{c}_1{}^{(\tilde{z})})_\xi$ coincides with the multiplication operator $\mathrm{M}[\exp(2\pi ik(\cdot))] \in \mathscr{L}(\mathcal{W}_1(\mathbb{T}))$ and thus the corresponding transition function is constant. This yields condition (c) of Definition A.1.1. □

**7.3.3 Proposition**
$\langle \mathfrak{C}_1 \rangle$ is a subbundle of $\langle \mathfrak{B}_1 \rangle$.

*Proof.*
Let $z \in \mathbb{C}\backslash\{0\}$. By Remark A.6.2 it suffices to show that the map $\mathcal{I}_\xi : \mathcal{W}_1(\mathbb{T}) \overset{((\mathfrak{c}_1{}^{(z)})_\xi)^{-1}}{\longrightarrow} (\mathfrak{C}_1)_\xi \hookrightarrow (\mathfrak{B}_1)_\xi \overset{(\mathfrak{b}_1{}^{(z)})_\xi}{\longrightarrow} \mathcal{W}_1[0,1]$ is analytically depending on $\xi \in \mathbb{B}_z$. By definition, for all $\xi \in \mathbb{B}_z$ $\mathcal{I}_\xi$ coincides with the embedding $\mathcal{W}_1(\mathbb{T}) \hookrightarrow \mathcal{W}_1[0,1]$ as a subspace. Thus obviously $[\xi \mapsto \mathcal{I}_\xi] \in A(\mathbb{B}_z, \mathscr{L}(\mathcal{W}_1(\mathbb{T}), \mathcal{W}_1[0,1]))$. □

**7.3.4 Remark**
Let ($A$-iv) hold.
Then Theorem 3.1.3 yields that $\mathcal{W}_0[0,1]$ and $\mathcal{W}_1(\mathbb{T})$ are isomorphic. Thus if $\{\phi^{(\lambda)} : \mathfrak{C}_{1|U_\lambda} \to U_\lambda \times B_\lambda\}_{\lambda\in\Lambda} \in \langle \mathfrak{C}_1 \rangle$ then by Proposition A.1.11 and Proposition A.1.13 there exists $\{\psi^{(\lambda)} : \mathfrak{C}_{1|U_\lambda} \to U_\lambda \times \mathcal{W}_0[0,1]\}_{\lambda\in\Lambda} \in \langle \mathfrak{C}_1 \rangle$.

## 7.4 The Bundles $\langle \mathfrak{B}'_1 \rangle$ and $\langle \mathfrak{C}'_1 \rangle$

We will now introduce bundles $\langle \mathfrak{B}'_1 \rangle$ and $\langle \mathfrak{C}'_1 \rangle$ that are defined analogously as $\langle \mathfrak{B}_1 \rangle$ and $\langle \mathfrak{C}_1 \rangle$, i. e. loosely speaking the fibers $\{z\} \times \mathcal{W}_1[0,1]$ and $\{z\} \times \mathcal{W}_{1,z}[0,1]$ are substituted by $\{z\} \times \mathcal{W}'_1[0,1]$ and $\{z\} \times \mathcal{W}'_{1,z}[0,1]$. For the sake of completeness, we explicitly give the following definitions. All statements follow analogously as in the situation of Section 7.3.

**7.4.1 Construction** (*The Bundle* $\langle \mathfrak{B}'_1 \rangle$)
We denote by $\langle \mathfrak{B}'_1 \rangle$ the bundle analogous to $\langle \mathfrak{B}_1 \rangle$, i. e. its total space is given $\mathfrak{B}'_1 := \mathbb{C}\backslash\{0\} \times \mathcal{W}'_1[0,1]$ and the defining trivialization is $\{\mathfrak{b}'_1 := \mathrm{Id} : \mathfrak{B}'_1 \to \mathbb{C}\backslash\{0\} \times \mathcal{W}'_1[0,1]\}$. A further trivialization that is equivalent to $\{\mathfrak{b}'_1\}$ is given by $\{\mathfrak{b}'^{(z)}_1\}_{z\in\mathbb{C}\backslash\{0\}}$ where $\mathfrak{b}'^{(z)}_1(\xi, f') := (\xi, \mathrm{M}[\xi^{(-\cdot)}_{(z)}]f')$ for all $z \in \mathbb{C}\backslash\{0\}$ and $(\xi, f') \in (\mathfrak{B}'_1)_{|\mathbb{B}_z}$. □

**7.4.2 Construction** (*The Bundle* $\langle \mathfrak{C}'_1 \rangle$)
We denote by $\langle \mathfrak{C}'_1 \rangle$ the bundle analogous to $\langle \mathfrak{C}_1 \rangle$, i. e. its total space is given by $\mathfrak{C}'_1 := \bigcup_{z\in\mathbb{C}\backslash\{0\}} \{z\} \times \mathcal{W}'_{1,z}[0,1]$ and the defining trivialization is $\{\mathfrak{c}'^{(z)}_1\}_{z\in\mathbb{C}\backslash\{0\}}$ where for each $z \in \mathbb{C}\backslash\{0\}$ $\mathfrak{c}'^{(z)}_1 : (\mathfrak{C}'_1)_{|\mathbb{B}_z} \to \mathbb{B}_z \times \mathcal{W}'_1(\mathbb{T})$ is the restriction of $\mathfrak{b}'^{(z)}_1$ to $(\mathfrak{C}'_1)_{|\mathbb{B}_z}$. (Again, for each $z \in \mathbb{C}\backslash\{0\}$ $(\mathfrak{C}'_1)_z$ is endowed with the Banach space structure of the linear subspace $\{z\} \times \mathcal{W}'_{1,z}[0,1]$ of $(\mathfrak{B}'_1)_z$.)

For all $z \in \mathbb{C}\backslash\{0\}$ and $\xi \in \mathbb{B}_z$ the induced map $(\mathfrak{c}_1'^{(z)})_\xi : (\mathfrak{C}_1')_\xi \to \mathcal{W}_1'(\mathbb{T})$ coincides with the multiplication operator $\mathrm{M}[\xi_{(z)}^{(-\cdot)}] \circ \cong \; \in \mathscr{L}((\mathfrak{C}_1')_\xi, \mathcal{W}_1'(\mathbb{T}))$, where $\cong$ denotes the natural identification of $(\mathfrak{C}_1')_\xi$ with $\mathcal{W}_{1,\xi}'[0,1]$, i. e. the isometric isomorphism $((\mathfrak{b}_1')_\xi)_{|(\mathfrak{C}_1')_\xi} \in \mathscr{L}((\mathfrak{C}_1')_\xi, \mathcal{W}_{1,\xi}'[0,1])$. □

**7.4.3 Proposition**
$\langle\mathfrak{C}_1'\rangle$ is a subbundle of $\langle\mathfrak{B}_1'\rangle$. □

**7.4.4 Remark**
Let ($A^*$-iv) hold.
If $\{\phi^{(\lambda)} : \mathfrak{C}_{1|U_\lambda}' \to U_\lambda \times B_\lambda\}_{\lambda\in\Lambda} \in \langle\mathfrak{C}_1'\rangle$ then there exists $\{\psi^{(\lambda)} : \mathfrak{C}_{1|U_\lambda}' \to U_\lambda \times \mathcal{W}_0'[0,1]\}_{\lambda\in\Lambda} \in \langle\mathfrak{C}_1'\rangle$.

## 7.5 The Bundle Homomorphisms $\mathfrak{L}$ and $\mathfrak{L}'$

**7.5.1 Definition and Proposition** ($\mathfrak{L}$)
Let ($A$-iv) and ($A$-v) hold.
We define $\mathfrak{L} : \mathfrak{C}_1 \to \mathfrak{B}_0$ by $\mathfrak{L}(z,f) := (z, \mathcal{L}_{\mathcal{W}_{1,z}[0,1]}f)$ for all $z \in \mathbb{C}\backslash\{0\}$ and $f \in \mathcal{W}_{1,z}[0,1]$.
Then $\mathfrak{L} : \langle\mathfrak{C}_1\rangle \to \langle\mathfrak{B}_0\rangle$ is an analytic Fredholm homomorphism.

*Proof.*
Let $z \in \mathbb{C}\backslash\{0\}$. Clearly, $(z, \mathcal{L}_{\mathcal{W}_{1,z}[0,1]}f) \in (\mathfrak{B}_0)_z$ for all $f \in \mathcal{W}_{1,z}[0,1]$ and thus $\mathfrak{L}$ is well-defined. Obviously, condition (a) of Definition A.4.1 holds. $\mathfrak{L}_z$ is given by $(\mathfrak{C}_1)_z \xrightarrow{\cong} \mathcal{W}_{1,z}[0,1] \xrightarrow{\mathcal{L}_{\mathcal{W}_{1,z}[0,1]}} \mathcal{W}_0[0,1] \xrightarrow{\cong} (\mathfrak{B}_0)_z$ where, by abuse of notation, $\cong$ denotes both the natural identification of $(\mathfrak{C}_1)_z$ with $\mathcal{W}_{1,z}[0,1]$ and of $\mathcal{W}_0[0,1]$ with $(\mathfrak{B}_0)_z$, resp.. Thus Corollary 3.1.9 implies that $\mathfrak{L}_z \in \mathscr{L}((\mathfrak{C}_1)_z, (\mathfrak{B}_0)_z)$ is a Fredholm operator. In particular, condition (b) of Definition A.4.1 holds. Then by Construction 7.3.2, for all $\xi \in \mathbb{B}_z$ the trivialized induced map $\mathcal{W}_1(\mathbb{T}) \xrightarrow{((\mathfrak{c}_1^{(z)})_\xi)^{-1}} (\mathfrak{C}_1)_\xi \xrightarrow{\mathfrak{L}_\xi} (\mathfrak{B}_0)_\xi \xrightarrow{(\mathfrak{b}_0)_\xi} \mathcal{W}_0[0,1]$ coincides with $\mathcal{L}_{\mathcal{W}_{1,\xi}[0,1]}\mathrm{M}[\xi_{(z)}^{(\cdot)}]$. By Fact 3.1.8 $\mathcal{L}_{\mathcal{W}_{1,\xi}[0,1]}\mathrm{M}[\xi_{(z)}^{(\cdot)}] = \mathrm{M}[\xi_{(z)}^{(\cdot)}](\mathcal{L}_{\mathcal{W}_1(\mathbb{T})} + \log_{(z)}\xi)$ for all $\xi \in \mathbb{B}_z$ and by Fact 1.9.1, Fact 1.5.10 and Fact 1.8.2 $[\xi \mapsto \mathrm{M}[\xi_{(z)}^{(\cdot)}](\mathcal{L}_{\mathcal{W}_1(\mathbb{T})} + \log_{(z)}\xi)] \in A(\mathbb{B}_z, \mathscr{L}(\mathcal{W}_1(\mathbb{T}), \mathcal{W}_0[0,1]))$. This yields condition (c) of Definition A.4.1. Therefore $\mathfrak{L} : \langle\mathfrak{C}_1\rangle \to \langle\mathfrak{B}_0\rangle$ is an analytic Fredholm homomorphism. □

**7.5.2 Corollary**
Let ($A$-iv) and ($A$-v) hold.
$\mathrm{S}(\mathfrak{L})$ and $\mathrm{CS}(\mathfrak{L})$ are analytic sets in $\mathbb{C}\setminus\{0\}$.

*Proof.*
Since $\mathfrak{L} : \langle\mathfrak{C}_1\rangle \to \langle\mathfrak{B}_0\rangle$ is an analytic Fredholm homomorphism, the assertion directly follows from [ZKKP75] § 5.0. □

Analogously, we obtain:

**7.5.3 Definition and Proposition** ($\mathfrak{L}'$)
Let ($A^*$-iv) and ($A^*$-v) hold.
We define $\mathfrak{L}' : \mathfrak{C}_1' \to \mathfrak{B}_0'$ by $\mathfrak{L}'(z, f) := (z, \mathcal{L}'_{\mathcal{W}'_{1,z}[0,1]} f)$ where $z \in \mathbb{C}\setminus\{0\}$ and $f \in \mathcal{W}'_{1,z}[0, 1]$.
Then $\mathfrak{L}' : \langle\mathfrak{C}_1'\rangle \to \langle\mathfrak{B}_0'\rangle$ is an analytic Fredholm homomorphism.

**7.5.4 Corollary**
Let ($A^*$-iv) and ($A^*$-v) hold.
$\mathrm{S}(\mathfrak{L}')$ and $\mathrm{CS}(\mathfrak{L}')$ are analytic sets in $\mathbb{C}\setminus\{0\}$.

## 7.6 The Floquet Transform

Analogously we will introduce two versions of the Floquet transform, one in the predual and one in the dual situation.

### The Floquet Transform on $\Phi_{0,\alpha}$ and $\Phi_{1,\alpha}$

**7.6.1 Definition**
For all $\alpha > 0$ we set $\mathbb{A}_\alpha := \{\, z \in \mathbb{C} : \exp(-\alpha) < |z| < \exp(\alpha) \,\}$.
Furthermore, we set $\mathbb{A}_\infty := \mathbb{C}\setminus\{0\}$.

For the convenience of the reader we state the following inclusions, cf. Remark A.2.9 and Remark 6.4.3.

**7.6.2 Remark**
Let $\alpha, \beta \in (0, \infty]$ with $\alpha \le \beta$.
Then

$$\mathbb{A}_\alpha \subset \mathbb{A}_\beta \subset \mathbb{C}\setminus\{0\},$$
$$\Gamma(\mathbb{C}\setminus\{0\}, \langle\mathfrak{B}_0\rangle) \subset \Gamma(\mathbb{A}_\beta, \langle\mathfrak{B}_0\rangle) \subset \Gamma(\mathbb{A}_\alpha, \langle\mathfrak{B}_0\rangle),$$

$\Gamma(\mathbb{C}\backslash\{0\}, \langle\mathfrak{C}_1\rangle) \subset \Gamma(\mathbb{A}_\beta, \langle\mathfrak{C}_1\rangle) \subset \Gamma(\mathbb{A}_\alpha, \langle\mathfrak{C}_1\rangle)$ and
$\Phi_j \subset \Phi_{j,\beta} \subset \Phi_{j,\alpha}$ for each $j \in \{0,1\}$.

For the rest of this section, let $\alpha \in (0, \infty]$.

**7.6.3 Remark and Convention**
By Proposition A.1.9 the restrictions $(\mathfrak{b}_0)_{|\mathbb{A}_\alpha}$, $(\mathfrak{b}_1{}^{(z)})_{|(\mathbb{B}_z \cap \mathbb{A}_\alpha)}$ and $(\mathfrak{c}_1{}^{(z)})_{|(\mathbb{B}_z \cap \mathbb{A}_\alpha)}$ (where $z \in \mathbb{A}_\alpha$) are trivializing maps for $\langle\mathfrak{B}_0\rangle$, $\langle\mathfrak{B}_1\rangle$ and $\langle\mathfrak{C}_1\rangle$, resp.. By abuse of notation we will use the same notation for the original maps and their restrictions within in this section.

**7.6.4 Construction** (*Floquet Transform on $\Phi_{0,\alpha}$*)
For all $\phi \in \Phi_{0,\alpha}$ and $k \in \mathbb{Z}$, by Remark 6.4.2 the Cauchy-Hadamard formula for Laurent series yields $[z \mapsto \sum_{k=-\infty}^{\infty} \phi_k z^k] \in A(\mathbb{A}_\alpha, \mathcal{W}_0[0,1])$, where $\phi_k := (\phi(\cdot - k))_{|[0,1]} \in \mathcal{W}_0[0,1]$. Thus

$$\mathcal{U}\phi := [z \mapsto ((\mathfrak{b}_0)_z)^{-1}(\sum_{k=-\infty}^{\infty} \phi_k z^k)] \in \Gamma(\mathbb{A}_\alpha, \langle\mathfrak{B}_0\rangle)$$

for all $\phi \in \Phi_{0,\alpha}$ and for clarity, we remark $(\mathcal{U}\phi)(z) = (z, \sum_{k=-\infty}^{\infty} \phi_k z^k)$ for all $\phi \in \Phi_{0,\alpha}$ and $z \in \mathbb{A}_\alpha$. Clearly, $\mathcal{U} : \Phi_{0,\alpha} \to \Gamma(\mathbb{A}_\alpha, \langle\mathfrak{B}_0\rangle)$ is linear and uniqueness of Laurent expansions implies that $\mathcal{U}$ is injective.

Conversely, let $s \in \Gamma(\mathbb{A}_\alpha, \langle\mathfrak{B}_0\rangle)$ and thus by Proposition A.2.4

$$[z \mapsto (\mathfrak{b}_0)_z(s(z))] \in A(\mathbb{A}_\alpha, \mathcal{W}_0[0,1])$$

can be expanded into a Laurent series with center 0 on $\mathbb{A}_\alpha$, say $\sum_{k=-\infty}^{\infty} s_k\, (\cdot)^k$ (with $s_k \in \mathcal{W}_0[0,1]$ for all $k \in \mathbb{Z}$). We set $\phi(t) := s_{-\lfloor t \rfloor}(t - \lfloor t \rfloor)$ for all $t \in \mathbb{R}$. Then[1] $\phi_k = s_k$ (pointwise on $[0,1)$ and hence in $\mathcal{W}_0[0,1]$) for all $k \in \mathbb{Z}$ and thus $\phi_{|[k,k+1]} \in \mathcal{W}_0[k,k+1]$. Hence by Fact 4.4.2 $\phi \in \mathcal{W}_{0,\text{loc}}(\mathbb{R})$ and the Cauchy-Hadamard formula[2] in combination with Remark 6.4.2 yields $\phi \in \Phi_{0,\alpha}$. Since $\mathcal{U}\phi = s$, we conclude that $\mathcal{U}$ is surjective.

Finally, we show that $\mathcal{U}$ is continuous: Let $\emptyset \neq K \subset\subset \mathbb{A}_\alpha$. Thus there is $a \in (0, \alpha)$ such that $\exp(-a) \leq |z| \leq \exp(a)$ for all $z \in K$. In particular, $|z|^k \leq \exp(a|k|)$ for all $z \in K$ and $k \in \mathbb{Z}$. Furthermore, let $\tilde{a} \in (a, \alpha)$. We calculate for all $\phi \in \Phi_{0,\alpha}$

$$\sup\nolimits_{z \in K} \|(\mathfrak{b}_0)_z(\mathcal{U}\phi)(z)\|_{\mathcal{W}_0[0,1]} =$$

[1] where $\phi_k := \phi(\cdot - k)$ as above
[2] applied to the coefficients $(s_k)_{k \in \mathbb{Z}}$

$$\sup_{z\in K} \| \sum_{k=-\infty}^{\infty} \phi_k z^k \|_{\mathcal{W}_0[0,1]} \leq$$

$$\sup_{z\in K} \sum_{k=-\infty}^{\infty} \|\phi_k\|_{\mathcal{W}_0[0,1]} |z^k| \leq$$

$$\sum_{k=-\infty}^{\infty} \|\phi\|_{\mathcal{W}_0[-k,-k+1]} \exp(a|k|) \leq$$

$$\big(\sup_{k\in\mathbb{Z}} \|\phi\|_{\mathcal{W}_0[k,k+1]} \exp(\tilde{a}|k|)\big)\big(\sum_{k=-\infty}^{\infty} \exp(-\tilde{a}|k|)\exp(a|k|)\big) \leq$$

$$\tfrac{\exp(\tilde{a}-a)+1}{\exp(\tilde{a}-a)-1} \cdot \gamma_0^{(\tilde{a})}(\phi).$$

Thus Proposition A.2.7 yields $\mathcal{U} \in \mathscr{L}(\Phi_{0,\alpha}, \Gamma(\mathbb{A}_\alpha, \langle\mathfrak{B}_0\rangle))$.

We resume:

$$\mathcal{U}: \begin{cases} \Phi_{0,\alpha} & \to & \Gamma(\mathbb{A}_\alpha, \langle\mathfrak{B}_0\rangle) \\ \phi & \mapsto & \mathcal{U}\phi & : \begin{cases} \mathbb{A}_\alpha & \to & \langle\mathfrak{B}_0\rangle \\ z & \mapsto & \big(z, \sum_{k=-\infty}^{\infty} \phi_k z^k\big) \end{cases} \end{cases}$$

is an isomorphism from $\Phi_{0,\alpha}$ to $\Gamma(\mathbb{A}_\alpha, \langle\mathfrak{B}_0\rangle)$ and is called *Floquet transform.*

△

**7.6.5 Remark**

In particular, $\mathcal{U}^* : (\Gamma(\mathbb{C}\backslash\{0\}, \langle\mathfrak{B}_0\rangle)))^* \to (\Phi_0)^*$ is an isomorphism of vector spaces.

We will now examine the restriction of $\mathcal{U}$ to $\Phi_1$.

**7.6.6 Construction** (*Floquet Transform on* $\Phi_{1,\alpha}$)

We will now show that $\mathcal{U}(\Phi_{1,\alpha}) = \Gamma(\mathbb{A}_\alpha, \langle\mathfrak{C}_1\rangle)$ by an application of Proposition A.2.4. Let $z \in \mathbb{A}_\alpha$. First, we remark that for all $\phi \in \Phi_{1,\alpha}$, analogously to Construction 7.6.4, $[\xi \mapsto (\xi, \sum_{k=-\infty}^{\infty} \phi_k \xi^k)] \in \Gamma(\mathbb{A}_\alpha, \langle\mathfrak{B}_1\rangle)$ where the series is to be understood as a $\mathcal{W}_1[0,1]$-valued Laurent series. Since $\mathcal{W}_1[0,1] \hookrightarrow \mathcal{W}_0[0,1]$ the so defined section coincides with $\mathcal{U}\phi$. In particular, by Proposition A.2.4 $\big[\xi \mapsto ({\mathfrak{b}_1}^{(z)})_\xi\big((\mathcal{U}\phi)(\xi)\big)\big] \in A(\mathbb{B}_z \cap \mathbb{A}_\alpha, \mathcal{W}_1[0,1])$. Next, for all $\phi \in \Phi_{1,\alpha}$ and $\xi \in \mathbb{A}_\alpha$ by Fact 2.4.1 $\xi\delta_0 \sum_{k=-\infty}^{\infty} \phi_k \xi^k = \sum_{k=-\infty}^{\infty} \delta_0 \phi_k \xi^{k+1} = \sum_{k=-\infty}^{\infty} \delta_1 \phi_{k+1} \xi^{k+1} = \delta_1 \sum_{k=-\infty}^{\infty} \phi_k \xi^k$ and thus $(\mathcal{U}\phi)(\xi) \in (\mathfrak{C}_1)_\xi$. Therefore $({\mathfrak{b}_1}^{(z)})_\xi\big((\mathcal{U}\phi)(\xi)\big) = ({\mathfrak{c}_1}^{(z)})_\xi\big((\mathcal{U}\phi)(\xi)\big) \in \mathcal{W}_1(\mathbb{T})$ for all $\xi \in \mathbb{B}_z \cap \mathbb{A}_\alpha$. By Fact 1.5.12 we obtain $\big[\xi \mapsto ({\mathfrak{c}_1}^{(z)})_\xi\big((\mathcal{U}\phi)(\xi)\big)\big] \in A(\mathbb{B}_z \cap \mathbb{A}_\alpha, \mathcal{W}_1(\mathbb{T}))$. Since obviously, $\nu \circ (\mathcal{U}\phi) = \mathrm{Id}_{\mathbb{A}_\alpha}$, Proposition A.2.4 yields $\mathcal{U}\phi \in \Gamma(\mathbb{A}_\alpha, \langle \mathfrak{C}_1 \rangle)$.

Conversely, let $s \in \Gamma(\mathbb{A}_\alpha, \langle \mathfrak{C}_1 \rangle)$. Then by Proposition 7.3.3 and Proposition A.7.1 $s \in \Gamma(\mathbb{A}_\alpha, \langle \mathfrak{B}_1 \rangle)$. Thus by Proposition A.2.4

$$[\xi \mapsto (\mathfrak{b}_1)_\xi(s(\xi))] \in A(\mathbb{A}_\alpha, \mathcal{W}_1[0,1])$$

can be expanded into a Laurent series with center 0 on $\mathbb{A}_\alpha$, say $\sum_{k=-\infty}^{\infty} s_k\, (\cdot)^k$ (with $s_k \in \mathcal{W}_1[0,1]$ for all $k \in \mathbb{Z}$). We set $\phi(t) := s_{-\lfloor t \rfloor}(t - \lfloor t \rfloor)$ for all $t \in \mathbb{R}$. Again since $\mathcal{W}_1[0,1] \hookrightarrow \mathcal{W}_0[0,1]$ we obtain $\phi = \mathcal{U}^{-1}s$ as in the previous construction. In particular,[3] $\phi_k = s_k$ (pointwise on $[0,1)$ and hence in $\mathcal{W}_1[0,1]$) for all $k \in \mathbb{Z}$ and thus $\phi_{|[k,k+1]} \in \mathcal{W}_1[k,k+1]$. Furthermore, for each $t = 0,1$ by Fact 2.4.1 and Fact 1.5.10 $[\xi \mapsto \delta_t\big((\mathfrak{b}_1)_\xi(s(\xi))\big)] \in A(\mathbb{A}_\alpha, X)$ and the Laurent expansion is given by $\sum_{k=-\infty}^{\infty} \delta_t(s_k)\,(\cdot)^k$. For every $\xi \in \mathbb{A}_\alpha$ $\sum_{k=-\infty}^{\infty} \delta_1(s_k)\xi^k = \delta_1\big((\mathfrak{b}_1)_\xi(s(\xi))\big) = \xi\delta_0\big((\mathfrak{b}_1)_\xi(s(\xi))\big) = \sum_{k=-\infty}^{\infty} \delta_0(s_k)\xi^{k+1}$ and thus uniqueness of the Laurent expansion yields $\delta_1 s_{k+1} = \delta_0 s_k$ for all $k \in \mathbb{Z}$. Thus[4] $\lim_{t \nearrow k} \phi(t) = \lim_{t \nearrow k} s_{-(k-1)}(t-(k-1)) = s_{-k+1}(1) = s_{-k}(0) = \phi(k)$ for all $k \in \mathbb{Z}$. By Fact 4.4.2 we obtain $\phi \in \mathcal{W}_{1,\mathrm{loc}}(\mathbb{R})$ and the Cauchy-Hadamard formula[5] in combination with Remark 6.4.2 yields $\phi \in \Phi_{1,\alpha}$. Thus indeed $\mathcal{U}(\Phi_{1,\alpha}) = \Gamma(\mathbb{A}_\alpha, \langle \mathfrak{C}_1 \rangle)$.

Finally, we will show that $\mathcal{U} \in \mathscr{L}(\Phi_{1,\alpha}, \Gamma(\mathbb{A}_\alpha, \langle \mathfrak{C}_1 \rangle))$ where, by abuse of notation, we denote by $\mathcal{U}$ the restriction of $\mathcal{U}$ to $\Phi_{1,\alpha}$. Let $z \in \mathbb{A}_\alpha$ and $\emptyset \neq K \subset\subset \mathbb{B}_z \cap \mathbb{A}_\alpha$.

Analogously as in Construction 7.6.4 there are $a, \tilde{a} \in (0,\alpha)$ with $a < \tilde{a}$ such that for all $\phi \in \Phi_{1,\alpha}$ $\sup_{\xi \in K} \|(\mathfrak{b}_1)_\xi(\mathcal{U}\phi)(\xi)\|_{\mathcal{W}_1[0,1]} \leq \frac{\exp(\tilde{a}-a)+1}{\exp(\tilde{a}-a)-1}\gamma_1^{(\tilde{a})}(\phi)$.

Furthermore

$$\sup_{\xi \in K} \|(\mathfrak{c}_1{}^{(z)})_\xi\|_{\mathscr{L}((\mathfrak{C}_1)_\xi, \mathcal{W}_1(\mathbb{T}))} = \sup_{\xi \in K} \|\mathrm{M}[\xi_{(z)}^{(-\cdot)}]\|_{\mathscr{L}(\mathcal{W}_{1,\xi}[0,1], \mathcal{W}_1(\mathbb{T}))} \leq$$

$$c := \sup_{\xi \in K} \|\xi_{(z)}^{(-\cdot)}\|_{C^1[0,1]} < \infty.$$

Therefore

$$\sup_{\xi \in K} \|(\mathfrak{c}_1{}^{(z)})_\xi(\mathcal{U}\phi)(\xi)\|_{\mathcal{W}_1(\mathbb{T})} \leq c \sup_{\xi \in K} \|(\mathcal{U}\phi)(\xi)\|_{(\mathfrak{C}_1)_\xi} =$$

$$c \sup_{\xi \in K} \|(\mathfrak{b}_1)_\xi(\mathcal{U}\phi)(\xi)\|_{\mathcal{W}_1[0,1]} \leq c \cdot \frac{\exp(\tilde{a}-a)+1}{\exp(\tilde{a}-a)-1} \cdot \gamma_1^{(\tilde{a})}(\phi) \text{ for all } \phi \in \Phi_{1,\alpha}.$$

Thus again Proposition A.2.7 yields $\mathcal{U} \in \mathscr{L}(\Phi_{1,\alpha}, \Gamma(\mathbb{A}_\alpha, \langle \mathfrak{C}_1 \rangle))$.

---

[3] where $\phi_k := \phi(\cdot - k)$ as above

[4] Here $\lim_{t \nearrow k}$ denotes the left-sided limit.

[5] applied to the coefficients $(s_k)_{k \in \mathbb{Z}}$

We resume: $\mathcal{U}$ is an isomorphism from $\Phi_{1,\alpha}$ to $\Gamma(\mathbb{A}_\alpha, \langle\mathfrak{C}_1\rangle)$. △

By a direct calculation using Remark 7.6.5 we obtain:

**7.6.7 Corollary**
$\mathfrak{L}_{\Gamma|\mathbb{A}_\alpha} = \mathcal{U} \circ \mathcal{L}_{\Phi_{1,\alpha}} \circ \mathcal{U}^{-1}$. In particular, $\mathfrak{L}_\Gamma = \mathcal{U} \circ \mathcal{L}_{\Phi_1} \circ \mathcal{U}^{-1}$.

### The Floquet Transform on $\Phi'_{0,\alpha}$ and $\Phi'_{1,\alpha}$

Throughout this section, let $\alpha \in (0, \infty]$.

Again, by applying all constructions of the previous section to the dual objects, we obtain the following results. By abuse of notation we denote by $\mathcal{U}$ also the corresponding map between the dual objects and its restrictions.

**7.6.8 Remark and Convention**
By Proposition A.1.9 the restrictions $(\mathfrak{b}'_0)_{|\mathbb{A}_\alpha}$, $(\mathfrak{b}'^{(z)}_1)_{|(\mathbb{B}_z\cap\mathbb{A}_\alpha)}$ and $(\mathfrak{c}'^{(z)}_1)_{|(\mathbb{B}_z\cap\mathbb{A}_\alpha)}$ (where $z \in \mathbb{A}_\alpha$) are trivializing maps for $\langle\mathfrak{B}'_0\rangle$, $\langle\mathfrak{B}'_1\rangle$ and $\langle\mathfrak{C}'_1\rangle$, resp.. By abuse of notation we will use the same notation for the original maps and their restrictions within in this section.

**7.6.9 Construction** (*Floquet Transform on* $\Phi'_{0,\alpha}$)
The following map, again called *Floquet transform*, is well-defined.

$$\mathcal{U} : \begin{cases} \Phi'_{0,\alpha} & \to & \Gamma(\mathbb{A}_\alpha, \langle\mathfrak{B}'_0\rangle) \\ \phi' & \mapsto & \mathcal{U}\phi' \end{cases} : \begin{cases} \mathbb{A}_\alpha & \to & \langle\mathfrak{B}'_0\rangle \\ z & \mapsto & \left(z, \sum\limits_{k=-\infty}^{\infty} \phi'_k z^k\right) \end{cases}$$

It is an isomorphism from $\Phi'_{0,\alpha}$ to $\Gamma(\mathbb{A}_\alpha, \langle\mathfrak{B}'_0\rangle)$. △

**7.6.10 Remark**
$\mathcal{U}^* : (\Gamma(\mathbb{C}\backslash\{0\}, \langle\mathfrak{B}'_0\rangle)))^* \to (\Phi'_0)^*$ is an isomorphism of vector spaces.

**7.6.11 Construction** (*Floquet Transform on* $\Phi'_{1,\alpha}$)
$\mathcal{U}$ is an isomorphism from $\Phi'_{1,\alpha}$ to $\Gamma(\mathbb{A}_\alpha, \langle\mathfrak{C}'_1\rangle)$. △

**7.6.12 Corollary**
$\mathfrak{L}'_{\Gamma|\mathbb{A}_\alpha} = \mathcal{U} \circ \mathcal{L}'_{\Phi'_{1,\alpha}} \circ \mathcal{U}^{-1}$. In particular, $\mathfrak{L}'_\Gamma = \mathcal{U} \circ \mathcal{L}'_{\Phi'_1} \circ \mathcal{U}^{-1}$.

Using Proposition 6.5.3 we additionally obtain:

**7.6.13 Corollary**
$\mathcal{U}^*(\operatorname{Coker}\mathfrak{L}'_\Gamma) = \operatorname{Coker}\mathcal{L}'_{\Phi'_1}$.

## 7.7 Form of Solution Functionals

Throughout this section let $z \in \mathbb{C}\backslash\{0\}$.

**7.7.1 Definition and Proposition**
If $n \in \mathbb{N}$ and for each $l = 1,\dots,n$ $p_l \in \mathbb{P}$, $1/z \in \Omega_l \overset{\circ}{\subset} \mathbb{C}\backslash\{0\}$ and $\sigma_l \in \Gamma(\Omega_l, \langle\mathfrak{B}_0\rangle)$ then $\big[s' \mapsto \sum_{l=1}^n \delta_{1/z}\big(p_l(\partial)\langle s', \sigma_l\rangle_{\Omega_l}\big)\big] \in (\Gamma(\mathbb{C}\backslash\{0\}, \langle\mathfrak{B}'_0\rangle))^*$.
The set of all functionals of that form (with all possible choices of $n$, $p_l$, $\Omega_l$ and $\sigma_l$) is denoted by $\mathcal{F}func_z$. $\mathcal{F}func_z$ is a linear subspace of $(\Gamma(\mathbb{C}\backslash\{0\}, \langle\mathfrak{B}'_0\rangle))^*$.
The set of all functionals in $\mathcal{F}func_z$ that have a representation of the above form where $p_l$ is a constant polynomial for each $l = 1,\dots,n$ is denoted by $\mathcal{B}func_z$.

*Proof.*
For each $l = 1,\dots,n$ by Remark A.2.9 and Proposition 7.2.3 $[s' \mapsto \langle s', \sigma_l\rangle_{\Omega_l}] \in \mathscr{L}(\Gamma(\mathbb{C}\backslash\{0\}, \langle\mathfrak{B}'_0\rangle), A(\Omega_l, \mathbb{C}))$, by Fact 1.5.7 $p_l(\partial) \in \mathscr{L}(A(\Omega_l, \mathbb{C}))$ and obviously $\delta_{1/z} \in (A(\Omega_l, \mathbb{C}))^*$. This yields $\big[s' \mapsto \sum_{l=1}^n \delta_{1/z}\big(p_l(\partial)\langle s', \sigma_l\rangle_{\Omega_l}\big)\big] \in (\Gamma(\mathbb{C}\backslash\{0\}, \langle\mathfrak{B}'_0\rangle))^*$. The vector space structure of $\mathcal{F}func_z$ is obvious. □

**7.7.2 Remark**
Let $\big[s' \mapsto \sum_{l=1}^n \delta_{1/z}\big(p_l(\partial)\langle s', \sigma_l\rangle_{\Omega_l}\big)\big] \in (\Gamma(\mathbb{C}\backslash\{0\}, \langle\mathfrak{B}'_0\rangle))^*$ (where $n$, $p_l$, $\Omega_l$ and $\sigma_l$ are defined as in Definition 7.7.1). Furthermore, let $1/z \in \Omega \overset{\circ}{\subset} \Omega_l$ for each $l = 1,\dots,n$. Then in view of Remark 7.2.4 obviously

$$\sum_{l=1}^n \delta_{1/z}\big(p_l(\partial)\langle s', \sigma_l\rangle_{\Omega}\big) = \sum_{l=1}^n \delta_{1/z}\big(p_l(\partial)\langle s', \sigma_l\rangle_{\Omega_l}\big)$$

for all $s' \in \Gamma(\mathbb{C}\backslash\{0\}, \langle\mathfrak{B}'_0\rangle)$. Thus we can always assume w. l. o. g., that for each $l = 1,\dots,n$ $\Omega_l$ can be chosen to be a "fixed but arbitrary small" neighborhood of $1/z$.

**7.7.3 Construction**
Let $f \in \mathcal{W}_0[0,1]$. Then $\mathfrak{s}_f := [\xi \mapsto ((\mathfrak{b}_0)_\xi)^{-1}(\xi^{(-\cdot)}_{(1/z)} f)] \in \Gamma(\mathbb{B}_{1/z}, \langle\mathfrak{B}_0\rangle)$.

*Proof.*
Clearly $p_0 \circ \mathfrak{s}_f = \mathrm{Id}_{\mathbb{B}_{1/z}}$. Fact 1.9.1 in combination with Fact 1.5.11 yields $[\xi \mapsto \xi^{(-\cdot)}_{(1/z)} f] \in A(\mathbb{B}_{1/z}, \mathcal{W}_0[0,1])$. Since $(\mathfrak{b}_0)_\xi(\mathfrak{s}_f(\xi)) = \xi^{(-\cdot)}_{(1/z)} f$ for all $\xi \in \mathbb{B}_{1/z}$ we conclude $\mathfrak{s}_f \in \Gamma(\mathbb{B}_{1/z}, \langle\mathfrak{B}_0\rangle)$. □

### 7.7.4 Proposition

$\mathcal{F}func_z$ is generated (as a vector space) by all functionals of the form $\left[s' \mapsto \delta_{1/z}\left(\partial^l \langle s', \mathfrak{s}_f\rangle_{\mathbb{B}_{1/z}}\right)\right] \in (\Gamma(\mathbb{C}\backslash\{0\}, \langle\mathfrak{B}_0'\rangle))^*$ where $l \in \mathbb{N}_0$ and $f \in \mathcal{W}_0[0,1]$.

*Proof.*

We will refer to functionals of the form mentioned in the statement as simple functionals. Obviously, $\mathcal{F}func_z$ is generated by all functionals of the form $S''_{l,\Omega,\sigma} := \left[s' \mapsto \delta_{1/z}\left(\partial^l \langle s', \sigma\rangle_\Omega\right)\right] \in (\Gamma(\mathbb{C}\backslash\{0\}, \langle\mathfrak{B}_0'\rangle))^*$ with $l \in \mathbb{N}_0$, $1/z \in \Omega \overset{\circ}{\subset} \mathbb{C}\backslash\{0\}$ and $\sigma \in (\Gamma(\Omega, \langle\mathfrak{B}_0\rangle)$. Thus it suffices to show that all functionals of that form are linear combinations of simple functionals.

Therefore let $l \in \mathbb{N}_0$, $1/z \in \Omega \overset{\circ}{\subset} \mathbb{C}\backslash\{0\}$ and $\sigma \in (\Gamma(\Omega, \langle\mathfrak{B}_0\rangle)$. By Remark 7.7.2 we can assume w.l.o.g. that $\Omega \subset \mathbb{B}_{1/z}$. Furthermore, there is $r > 0$ such that $\overline{B_{\mathbb{C}}(1/z, r)} \subset \Omega$ and we set $\widetilde{\Omega} := B_{\mathbb{C}}(1/z, r)$. Thus by Fact 1.5.13 analytic functions on $\Omega$ can be expanded into power series about $1/z$ on $\widetilde{\Omega}$.

By Proposition A.2.4 $[\xi \mapsto (\mathfrak{b}_0)_\xi(\sigma(\xi))] \in A(\Omega, \mathcal{W}_0[0,1])$ and therefore by Fact 1.5.11 $[\xi \mapsto \xi_{(1/z)}^{(\cdot)}(\mathfrak{b}_0)_\xi(\sigma(\xi))] \in A(\Omega, \mathcal{W}_0[0,1])$. By Fact 1.9.1 $[\xi \mapsto \xi_{(1/z)}^{(-\cdot)}] \in A(\Omega, C[0,1])$. Again by Proposition A.2.4 $[\xi \mapsto (\mathfrak{b}_0')_\xi(s'(\xi))] \in A(\Omega, \mathcal{W}_0'[0,1])$ for all $s' \in \Gamma(\mathbb{C}\backslash\{0\}, \langle\mathfrak{B}_0'\rangle)$.

We denote the power series expansion about $1/z$ on $\widetilde{\Omega}$ of

$$\xi \mapsto \xi_{(1/z)}^{(\cdot)}(\mathfrak{b}_0)_\xi(\sigma(\xi)),$$

$$\xi \mapsto \xi_{(1/z)}^{(-\cdot)} \text{ and}$$

$$\xi \mapsto (\mathfrak{b}_0')_\xi(s'(\xi)),$$

where $s' \in \Gamma(\mathbb{C}\backslash\{0\}, \langle\mathfrak{B}_0'\rangle)$, by

$$\sum_{\alpha=0}^{\infty} f_\alpha(\cdot - 1/z)^\alpha,$$

$$\sum_{\beta=0}^{\infty} c_\beta(\cdot - 1/z)^\beta \text{ and}$$

$$\sum_{\gamma=0}^{\infty} f_\gamma'(\cdot - 1/z)^\gamma, \text{ resp..}$$

Note that $\delta_{1/z}(\partial^n[\Omega \ni \xi \mapsto (\xi - 1/z)^m]) = n!\delta_{n,m}$ for all $n, m \in \mathbb{N}_0$. Thus for every $f \in \mathcal{W}_0[0,1]$ and every $n \in \mathbb{N}_0$, linearity and continuity of the involved operators in combination with Fact 1.5.13 yield

$$\delta_{1/z}\big(\partial^n\langle s',\mathfrak{s}_f\rangle_{\widetilde{\Omega}}\big) =$$
$$\delta_{1/z}\big(\partial^n\big[\widetilde{\Omega}\ni\xi\mapsto\langle(\mathfrak{b}_0')_\xi(s'(\xi)),(\mathfrak{b}_0)_\xi(((\mathfrak{b}_0)_\xi)^{-1}(\xi_{(1/z)}^{(-\cdot)}f))\rangle_W\big]\big) =$$
$$\sum_{\gamma=0}^{\infty}\sum_{\beta=0}^{\infty}\delta_{1/z}\big(\partial^n\big[\widetilde{\Omega}\ni\xi\mapsto(\xi-1/z)^{\gamma+\beta}\langle f_\gamma',c_\beta f\rangle_W\big]\big) =$$
$$n!\sum_{\gamma+\beta=n}\langle f_\gamma',c_\beta f\rangle_W.$$

Similarly, we obtain for all $s'\in\Gamma(\mathbb{C}\backslash\{0\},\langle\mathfrak{B}_0'\rangle)$

$$S''_{l,\Omega,\sigma}(s') = \delta_{1/z}\big(\partial^l\langle s',\sigma\rangle_{\widetilde{\Omega}}\big) =$$
$$\delta_{1/z}\big(\partial^l\big[\widetilde{\Omega}\ni\xi\mapsto\langle(\mathfrak{b}_0')_\xi(s'(\xi)),\xi_{(1/z)}^{(-\cdot)}\xi_{(1/z)}^{(\cdot)}(\mathfrak{b}_0)_\xi(\sigma(\xi))\rangle_W\big]\big) =$$
$$\sum_{\gamma=0}^{\infty}\sum_{\beta=0}^{\infty}\sum_{\alpha=0}^{\infty}\delta_{1/z}\big(\partial^l\big[\widetilde{\Omega}\ni\xi\mapsto(\xi-1/z)^{\gamma+\beta+\alpha}\langle f_\gamma',c_\beta f_\alpha\rangle_W\big]\big) =$$
$$\sum_{\gamma+\beta+\alpha=l} l!\langle f_\gamma',c_\beta f_\alpha\rangle_W =$$
$$\sum_{\alpha=0}^{l} l!\sum_{\gamma+\beta=l-\alpha}\langle f_\gamma',c_\beta f_\alpha\rangle_W$$
$$\sum_{\alpha=0}^{l}\tfrac{l!}{(l-\alpha)!}\delta_{1/z}\big(\partial^{(l-\alpha)}\langle s',\mathfrak{s}_{f_\alpha}\rangle_{\widetilde{\Omega}}\big),$$

where the last step follows from applying the first calculation to $f_\alpha$ instead of $f$.

We conclude that $S''_{l,\widetilde{\Omega},\sigma}$ is a linear combination of simple functionals. By Remark 7.7.2 $S''_{l,\widetilde{\Omega},\sigma}$ coincides with $S''_{l,\Omega,\sigma}$ as a functional on $\Gamma(\mathbb{C}\backslash\{0\},\langle\mathfrak{B}_0'\rangle)$. This finishes the proof. □

**7.7.5 Proposition**

Let $s''\in(\Gamma(\mathbb{C}\backslash\{0\},\langle\mathfrak{B}_0'\rangle))^*$.

Then the following are all equivalent:

(a) $s''\in\mathcal{B}func_z$,

(b) $s''$ is of the form $s'\mapsto\delta_{1/z}\langle s',\sigma\rangle_\Omega$ where $1/z\in\Omega\overset{\circ}{\subset}\mathbb{C}\backslash\{0\}$ and $\sigma\in\Gamma(\Omega,\langle\mathfrak{B}_0\rangle)$,

(c) $s''$ is of the form $s'\mapsto\langle s'(1/z),\mathfrak{f}\rangle_{1/z}$ where $\mathfrak{f}\in(\mathfrak{B}_0)_{1/z}$,

(d) $s''$ is of the form $s' \mapsto \delta_{1/z}\langle s', \mathfrak{s}_f\rangle_{\mathbb{B}_{1/z}}$ where $f \in \mathcal{W}_0[0,1]$.

*Proof.*

"(a)⇒(b)": Let $s'' \in \mathcal{B}func_z$. Then by definition

$$s'' = \big[s' \mapsto \textstyle\sum_{l=1}^n \delta_{1/z}\big(p_l\langle s', \sigma_l\rangle_{\Omega_l}\big)\big]$$

with $n \in \mathbb{N}$ and for each $l = 1, \dots, n$ $p_l \in \mathbb{C}$, $1/z \in \Omega_l \overset{\circ}{\subset} \mathbb{C}\backslash\{0\}$ and $\sigma_l \in \Gamma(\Omega_l, \langle\mathfrak{B}_0\rangle)$. By Remark 7.7.2 and Proposition A.2.5 we obtain $s'' = [s' \mapsto \delta_{1/z}\langle s', \sigma\rangle_\Omega]$ where $\Omega := \bigcap_{k=1}^n \Omega_l$ and $\sigma := \sum_{l=1}^n p_l\sigma_l$.

"(b)⇒(c)": Let $s''$ be of the form $s' \mapsto \delta_{1/z}\langle s', \sigma\rangle_\Omega$ where $1/z \in \Omega \overset{\circ}{\subset} \mathbb{C}\backslash\{0\}$ and $\sigma \in \Gamma(\Omega, \langle\mathfrak{B}_0\rangle)$. Thus $s'' = [s' \mapsto \langle s'(1/z), \sigma(1/z)\rangle_{1/z}]$. Since $\mathfrak{f} := \sigma(1/z) \in (\mathfrak{B}_0)_{1/z}$, $s''$ has indeed the required form.

"(c)⇒(d)": Let $s''$ be of the form $s' \mapsto \langle s'(1/z), \mathfrak{f}\rangle_{1/z}$ where $\mathfrak{f} \in (\mathfrak{B}_0)_{1/z}$ Then $(\mathfrak{b}_0)_{(1/z)}\mathfrak{f} \in \mathcal{W}_0[0,1]$ and thus $f := (1/z)^{(\cdot)}_{(1/z)}(\mathfrak{b}_0)_{(1/z)}\mathfrak{f} \in \mathcal{W}_0[0,1]$. Thus by definition $\mathfrak{s}_f(1/z) = ((\mathfrak{b}_0)_{(1/z)})^{-1}((1/z)^{(-\cdot)}_{(1/z)}f) = \mathfrak{f}$ and $[s' \mapsto \delta_{1/z}\langle s', \mathfrak{s}_f\rangle_{\mathbb{B}_{1/z}}] = [s' \mapsto \langle s'(1/z), \mathfrak{s}_f(1/z)\rangle_{1/z}] = s''$. Therefore $s''$ is of the required form.

"(d)⇒(a)": This is a direct consequence of Construction 7.7.3. □

### 7.7.6 Proposition

$\mathrm{F}(\mathcal{F}form_z) = \mathcal{U}^*(\mathcal{F}func_z)$ and $\mathrm{F}(\mathcal{B}form_z) = \mathcal{U}^*(\mathcal{B}func_z)$.

*Proof.*

Let $s'' \in \mathcal{F}func_z$. Let us assume for the moment that $s''$ has the form described in Proposition 7.7.4, i.e. there are $l \in \mathbb{N}_0$ and $f \in \mathcal{W}_0[0,1]$ such that $s'' = \big[\Gamma(\mathbb{C}\backslash\{0\}, \langle\mathfrak{B}_0'\rangle) \ni s' \mapsto \delta_{1/z}\big(\partial^l\langle s', \mathfrak{s}_f\rangle_{\mathbb{B}_{1/z}}\big)\big]$. Then for all $\phi' \in C_c^\infty(\mathbb{R}, X^*)$

$$(\mathcal{U}^*s'')(\phi') = s''(\mathcal{U}\phi') = \delta_{1/z}\big(\partial^l\langle\mathcal{U}\phi', \mathfrak{s}_f\rangle_{\mathbb{B}_{1/z}}\big) =$$

$$\delta_{1/z}\big(\partial^l\big[\mathbb{B}_{1/z} \ni \xi \mapsto \big\langle(\mathfrak{b}_0')_\xi\big((\mathcal{U}\phi')(\xi)\big), (\mathfrak{b}_0)_\xi\big(\mathfrak{s}_f(\xi)\big)\big\rangle_{\mathcal{W}}\big]\big) =$$

$$\delta_{1/z}\big(\partial^l\big[\mathbb{B}_{1/z} \ni \xi \mapsto \int_0^1 \big\langle \sum_{k=-\infty}^{\infty} \xi^k\phi'(t-k), \xi^{-t}_{(1/z)}f(t)\big\rangle_X\, dt\big]\big) =$$

$$\delta_{1/z}\big(\partial^l\big[\mathbb{B}_{1/z} \ni \xi \mapsto \sum_{k=-\infty}^{\infty} \int_0^1 \xi^{k-t}_{(1/z)}\langle\phi'(t-k), f(t)\rangle_X\, dt\big]\big) =$$

$$\delta_{1/z}\big(\partial^l\big[\mathbb{B}_{1/z} \ni \xi \mapsto \sum_{k=-\infty}^{\infty} \int_{-k}^{-k+1} \xi^{-\tau}_{(1/z)}\langle\phi'(\tau), f(\tau+k)\rangle_X\, d\tau\big]\big) =$$

$$\delta_{1/z}\big(\partial^l\big[\mathbb{B}_{1/z} \ni \xi \mapsto \int_{\mathbb{R}} \xi^{-\tau}_{(1/z)}\langle\phi'(\tau), \mathrm{E}_{\mathbb{T}}f(\tau)\rangle_X\, d\tau\big]\big) =$$

$$\delta_{1/z}\big[\mathbb{B}_{1/z} \ni \xi \mapsto \int_{\mathbb{R}} (\partial^l[\mathbb{B}_{1/z} \ni \tilde{\xi} \mapsto \tilde{\xi}^{-\tau}_{(1/z)}])(\xi)\,\langle \phi'(\tau), \mathrm{E}_{\mathbb{T}} f(\tau)\rangle_X \,d\tau\big] =$$

$$\delta_{1/z}\big[\mathbb{B}_{1/z} \ni \xi \mapsto \int_{\mathbb{R}} p_l(\tau)\xi^{-\tau-l}_{(1/z)}\langle \phi'(\tau), \mathrm{E}_{\mathbb{T}} f(\tau)\rangle_X \,d\tau\big] =$$

$$\int_{\mathbb{R}} \langle \phi'(\tau), (1/z)^{-\tau}_{(1/z)} p_l(\tau)(1/z)^{-l}\mathrm{E}_{\mathbb{T}} f(\tau)\rangle_X \,d\tau =$$

$$\int_{\mathbb{R}} \langle \phi'(\tau), \exp(-\tau \log_{(1/z)}(1/z)) p_l(\tau) z^l \mathrm{E}_{\mathbb{T}} f(\tau)\rangle_X \,d\tau =$$

$$\mathrm{F}[\mathbb{R} \ni t \mapsto \exp(t\lambda_z) p_l(t) z^l \mathrm{E}_{\mathbb{T}} f(t)](\phi'),$$

where[6] $p_l(\tau) := \prod_{m=0}^{l-1}(-\tau - m)$ for all $\tau \in \mathbb{R}$ and $\lambda_z := -\log_{(1/z)}(1/z)$.
Density of $C_c^\infty(\mathbb{R}, X^*)$ in $\Phi_0'$ thus yields

$$\mathcal{U}^* s'' = \mathrm{F}[\mathbb{R} \ni t \mapsto \exp(t\lambda_z) p_l(t) z^l \mathrm{E}_{\mathbb{T}} f(t)].$$

We remark that since $\exp(\lambda_z) = z$

$$u_{\lambda_z,l,p_l,f} := [\mathbb{R} \ni t \mapsto \exp(t\lambda_z) p_l(t) z^l \mathrm{E}_{\mathbb{T}} f(t)] \in \mathcal{F}form_z.$$

Now, let $s'' \in \mathcal{F}func_z$ be arbitrary. By Proposition 7.7.4 $s''$ can be represented as a linear combination of functionals of the above form. Hence $\mathcal{U}^*(\mathcal{F}func_z) \subset \mathrm{F}(\mathcal{F}form_z)$.

Conversely, let $u \in \mathcal{F}form_z$. Obviously $\{p_l\}_{l\in\mathbb{N}_0}$ is a basis for the space of polynomials. In combination with Proposition 6.1.6 we obtain that $u$ can be written as linear combination of functions $t \mapsto \exp(\lambda_z t) p_l(t) g_l(t)$ a. e. on $\mathbb{R}$ with $l \in \mathbb{N}_0$ and $g_l \in L_p(\mathbb{T}, X)$. Since $[t \mapsto \exp(\lambda_z t) p_l(t) g_l(t)] = u_{\lambda_z,l,p_l,\tilde{g}_l}$ with $\tilde{g}_l := z^{-l}(g_l)_{|[0,1]}$ the above calculation yields $\mathrm{F}u \in \mathcal{U}^*(\mathcal{F}func_z)$, hence $\mathrm{F}(\mathcal{F}form_z) \subset \mathcal{U}^*(\mathcal{F}func_z)$.

Finally, if $s'' \in \mathcal{B}func_z$ or $u \in \mathcal{B}form_z$, then by Proposition 7.7.5 and Proposition 6.1.6, resp., $s''$ and $u$ have the corresponding representations with $l = 0$. This directly yields $\mathrm{F}(\mathcal{B}form_z) = \mathcal{U}^*(\mathcal{B}func_z)$. □

### 7.7.7 Proposition

For each $z \in \mathcal{F}set$

$$\mathrm{F}(\mathcal{F}sol_z) = \mathcal{U}^*(\mathcal{F}func_z \cap \operatorname{Coker} \mathfrak{L}'_\Gamma) \setminus \{0\} \text{ and}$$
$$\mathrm{F}(\mathcal{B}sol_z) = \mathcal{U}^*(\mathcal{B}func_z \cap \operatorname{Coker} \mathfrak{L}'_\Gamma) \setminus \{0\} \text{ holds.}$$

[6] Here, we set $\prod_{m=0}^{-1}(-\tau - m) := 1$.

*Proof.*
This is a direct consequence of Proposition 7.7.6, Proposition 6.5.3 and Corollary 7.6.13. □

## 7.8 Characterization of Floquet Exponents

**7.8.1 Proposition**
Let ($A$-iv) and ($A^*$-v) hold.
Then $\mathcal{B}set = \mathcal{F}set = (\mathrm{CS}(\mathfrak{L}'))^{-1} = \mathrm{S}(\mathfrak{L})$.

*Proof.*
We will show $\mathcal{B}set \subset \mathcal{F}set \subset (\mathrm{CS}(\mathfrak{L}'))^{-1} \subset \mathrm{S}(\mathfrak{L}) \subset \mathcal{B}set$.

"$\mathcal{B}set \subset \mathcal{F}set$":

The definition of $\mathcal{B}set$ directly yields $\mathcal{B}set \subset \mathcal{F}set$.

"$\mathcal{F}set \subset (\mathrm{CS}(\mathfrak{L}'))^{-1}$":

Let $z \in \mathcal{F}set$.

We will prove the statement by contradiction. So, assume that $z^{-1} \notin \mathrm{CS}(\mathfrak{L}')$. By Corollary 7.5.4 $\mathrm{CS}(\mathfrak{L}')$ is an analytic set in $\mathbb{C}\backslash\{0\}$. In particular, $\mathrm{CS}(\mathfrak{L}')$ is closed in $\mathbb{C}\backslash\{0\}$. Therefore there exists a neighborhood $\Omega$ of $z^{-1}$ such that $\Omega \cap \mathrm{CS}(\mathfrak{L}') = \emptyset$. W. l. o. g. we can assume that $\Omega$ is connected and that[7] there are trivializing maps $\mathfrak{c}' : (\mathfrak{C}_1')_{|\Omega} \to \Omega \times \mathcal{W}_0'[0,1]$ for $\langle \mathfrak{C}_1' \rangle$ and $\mathfrak{b}' : (\mathfrak{B}_0')_{|\Omega} \to \Omega \times \mathcal{W}_0'[0,1]$ for $\langle \mathfrak{B}_0' \rangle$, resp..

Then the induced homomorphism $\mathfrak{L}'_{\Gamma|\Omega} : \Gamma(\Omega, \langle \mathfrak{C}_1' \rangle) \to \Gamma(\Omega, \langle \mathfrak{B}_0' \rangle)$ is surjective: First, we remark that for all $\xi \in \Omega$ by Fact 1.2.2 $\{0\} = \mathrm{Coker}\, \mathfrak{L}'_\xi \cong (\mathfrak{B}_0')_\xi / (\mathrm{Range}\, \mathfrak{L}'_\xi)$ and therefore $\mathfrak{L}'_\xi : (\mathfrak{C}_1')_\xi \to (\mathfrak{B}_0')_\xi$ is surjective. We define $L' : \Omega \to \mathscr{L}(\mathcal{W}_0'[0,1])$ by $L'(\xi) := \mathfrak{b}'_\xi \circ \mathfrak{L}'_\xi \circ (\mathfrak{c}'_\xi)^{-1}$ for all $\xi \in \Omega$. Thus, $L'(\xi) \in \mathscr{L}(\mathcal{W}_0'[0,1])$ is surjective for each $\xi \in \Omega$. Since $\mathfrak{L}'_\xi$ is a Fredholm operator, $L'(\xi)$ has a finite and thus complemented kernel. Therefore $L'(\xi)$ is right-invertible for each $\xi \in \Omega$, cf. [Heu92] § 26 Aufgabe 4. Furthermore, by Proposition 7.5.3 and Proposition A.4.3 $L' \in A(\Omega, \mathscr{L}(\mathcal{W}_0'[0,1]))$. By the central theorem of [All67][8] there is $R' \in A(\Omega, \mathscr{L}(\mathcal{W}_0'[0,1]))$ such that $R'(\xi)$ is a

[7] Cf. Proposition A.1.10 and Remark 7.4.4

[8] Actually, the theorem only yields the existence of a right inverse on a "little smaller" (but still connected and open) set than $\Omega$. However, if need be, we can choose $\Omega$ to be that smaller domain and therefore assume w. l. o. g. the result holds as cited.

right inverse of $L'(\xi)$ for all $\xi \in \Omega$. Now, let $s' \in \Gamma(\Omega, \langle\mathfrak{B}_0'\rangle)$. For all $\xi \in \Omega$, we set $\sigma'(\xi) := \left((\mathfrak{c}_\xi')^{-1} R'(\xi)\mathfrak{b}_\xi'\right)\left(s'(\xi)\right)$. Since $[\xi \mapsto \mathfrak{b}_\xi'\left(s'(\xi)\right)] \in A(\Omega, \mathcal{W}_0'[0,1])$, by Fact 1.5.10 we obtain $[\xi \mapsto \mathfrak{c}_\xi'\left(\sigma'(\xi)\right)] \in A(\Omega, \mathcal{W}_0'[0,1])$. Clearly, $p_1 \circ \sigma' = \mathrm{Id}_\Omega$ and thus $\sigma' \in \Gamma(\Omega, \langle\mathfrak{C}_1'\rangle)$. Furthermore, $\mathfrak{L}'\left(\sigma'(\xi)\right) = \mathfrak{L}_\xi'\left(\sigma'(\xi)\right) = \left((\mathfrak{b}_\xi')^{-1}\mathfrak{b}_\xi'\mathfrak{L}_\xi'(\mathfrak{c}_\xi')^{-1}R'(\xi)\mathfrak{b}_\xi'\right)\left(s'(\xi)\right) = \left((\mathfrak{b}_\xi')^{-1} \circ L'(\xi) \circ R'(\xi) \circ \mathfrak{b}_\xi' s'\right)(\xi) = s'(\xi)$ for all $\xi \in \Omega$. Thus $\mathfrak{L}_{\Gamma|\Omega}'\sigma' = s'$. This proves that $\mathfrak{L}_{\Gamma|\Omega}'$ is surjective.

Proposition 7.7.7 yields that there is $0 \neq S'' \in \mathcal{F}func_z \cap \operatorname{Coker} \mathfrak{L}_\Gamma'$. Again, by choosing $\Omega$ smaller, if need be, by Remark 7.7.2 we can assume w. l. o. g. that $S'' = \left[\Gamma(\mathbb{C}\backslash\{0\}, \langle\mathfrak{B}_0'\rangle) \ni s' \mapsto \sum_{l=1}^n \delta_{1/z}\left(p_l(\partial)\langle s', \sigma_l\rangle_\Omega\right)\right]$, where $n \in \mathbb{N}$ and for each $l = 1, \dots, n$ $p_l \in \mathbb{P}$ and $\sigma_l \in (\Gamma(\Omega, \langle\mathfrak{B}_0\rangle)$.

We will now show the contradiction $S'' = 0$: Let $s_0' \in \Gamma(\mathbb{C}\backslash\{0\}, \langle\mathfrak{B}_0'\rangle)$. Then there exists $\sigma_0' \in \Gamma(\Omega, \langle\mathfrak{C}_1'\rangle)$ such that $\mathfrak{L}_{\Gamma|\Omega}'\sigma_0' = (s_0')_{|\Omega}$.

Then by [Lei78] Theorem 2.3 (iv) there exists[9] $(\sigma_n')_{n\in\mathbb{N}} \subset \Gamma(\mathbb{C}\backslash\{0\}, \langle\mathfrak{C}_1'\rangle)$ with $(\sigma_n')_{|\Omega} \overset{n\to\infty}{\longrightarrow} \sigma_0'$ in $\Gamma(\Omega, \langle\mathfrak{C}_1'\rangle)$. Thus $\mathfrak{L}_{\Gamma|\Omega}'\left((\sigma_n')_{|\Omega}\right) \overset{n\to\infty}{\longrightarrow} (s_0')_{|\Omega}$ in $\Gamma(\Omega, \langle\mathfrak{B}_0'\rangle)$. By definition, $S''(\mathfrak{L}_\Gamma'\sigma_n') = 0$ for all $n \in \mathbb{N}$. On the other hand, by Proposition 7.7.1 and by continuity of the involved operators[10]

$$
\begin{aligned}
&S''(\mathfrak{L}_\Gamma'\sigma_n') = \\
&\sum_{l=1}^n \delta_{1/z}\left(p_l(\partial)\langle\mathfrak{L}_\Gamma'\sigma_n', \sigma_l\rangle_\Omega\right) = \\
&\sum_{l=1}^n \delta_{1/z}\left(p_l(\partial)\langle\mathfrak{L}_{\Gamma|\Omega}'\left((\sigma_n')_{|\Omega}\right), \sigma_l\rangle_\Omega\right) \overset{n\to\infty}{\longrightarrow} \\
&\sum_{l=1}^n \delta_{1/z}\left(p_l(\partial)\langle(s_0')_{|\Omega}, \sigma_l\rangle_\Omega\right) = \\
&\sum_{l=1}^n \delta_{1/z}\left(p_l(\partial)\langle s_0', \sigma_l\rangle_\Omega\right) = \\
&S''(s_0').
\end{aligned}
$$

This yields $S''(s_0') = 0$ and therefore $S'' = 0$.

"$(\mathrm{CS}(\mathfrak{L}'))^{-1} \subset \mathrm{S}(\mathfrak{L})$":

Let $z$ in $(\mathrm{CS}(\mathfrak{L}'))^{-1}$. Then by Remark 7.2.2 there exists $0 \neq f \in (\mathfrak{B}_0)_{1/z}$ such

---

[9] We note that $\Omega$ is (as every domain in $\mathbb{C}$) holomorphically convex (cf. [GR65] definition VII.A.2 and the subsequent example (3)) and that by Fact B.1.29 $\mathcal{O}^{\langle\mathfrak{C}_1'\rangle}(\mathbb{C}\backslash\{0\})$ is a BCAF sheaf.

[10] Namely, $\langle\cdot,\cdot\rangle_\Omega \in \mathscr{L}(\Gamma(\Omega, \langle\mathfrak{B}_0'\rangle); \Gamma(\Omega, \langle\mathfrak{B}_0\rangle), A(\Omega, \mathbb{C}))$ by Proposition 7.2.3, $\partial \in \mathscr{L}(A(\Omega, \mathbb{C}))$ by Fact 1.5.7 and $\delta_{1/z} \in (A(\Omega, \mathbb{C}))^*$ by Fact 1.5.8.

that $\langle \mathfrak{L}'_{1/z} f', f\rangle_{1/z} = 0$ for all $f' \in (\mathfrak{C}'_1)_{1/z}$. By Proposition 7.7.5 $S'' := [s' \mapsto \langle s'(1/z), f\rangle_{1/z}] \in \mathcal{B}func_z$. Clearly, $S'' \neq 0$. Since for all $s' \in \Gamma(\mathbb{C}\backslash\{0\}, \langle \mathfrak{C}'_1\rangle)$ $S''(\mathfrak{L}'_\Gamma s') = \langle (\mathfrak{L}' \circ s')(1/z), f\rangle_{1/z} = \langle \mathfrak{L}'_{1/z}(s'(1/z)), f\rangle_{1/z} = 0$, $S'' \in \operatorname{Coker} \mathfrak{L}'_\Gamma$. Thus by Proposition 7.7.7 there is $0 \neq u \in \mathcal{B}sol_z$. By Fact 6.1.5 $u$ is $z$-quasiperiodic and thus Theorem 5.1.7 yields $u \in \mathcal{W}_{1,z}[0,1]$ and $\mathcal{L}_{\mathcal{W}_{1,z}[0,1]} u = 0$. Hence[11] $\mathfrak{L}_z(z, u) = (z, \mathcal{L}_{\mathcal{W}_{1,z}[0,1]} u) = 0$ and $z \in \mathrm{S}(\mathfrak{L})$.

"$\mathrm{S}(\mathfrak{L}) \subset \mathcal{B}set$":

Let $z \in \mathrm{S}(\mathfrak{L})$. Then there is $\tilde{f} \in \operatorname{Ker} \mathfrak{L}_z$, i.e. there is $f \in \mathcal{W}_{1,z}[0,1]$ such that $\tilde{f} = (z, f)$ and $\mathcal{L}_{\mathcal{W}_{1,z}[0,1]} f = 0$. Then by Fact 4.4.5 $\mathrm{E}_z f \in \mathcal{W}_{1,\mathrm{loc}}(\mathbb{R})$ and $\mathcal{L}_{\mathcal{W}_{1,\mathrm{loc}}(\mathbb{R})}(\mathrm{E}_z f) = 0$. Hence Proposition 4.5.3 yields that $\mathrm{E}_z f$ is a solution. Furthermore, by Fact 6.1.5 $\mathrm{E}_z f \in \mathcal{B}form_z$ and thus $\mathrm{E}_z f \in \mathcal{B}sol_z$. We conclude $z \in \mathcal{B}set$. □

[11]Here, 0 denotes the zero vector in $(\mathfrak{B}_0)_z$.

# Chapter 8

# The Superposition Result

For the convenience of the reader we recall the most important notions.

We examine solutions (cf. Definition 4.5.2) to the equation

$$u'(t) + A_t u(t) = 0 \qquad (t \in \mathbb{R}) \tag{E}$$

where the following conditions hold for the family $(A_t : X \supset D(A_t) \to X)_{t\in\mathbb{T}}$ of closed operators depending *periodically* on $t$.

($A$-i) There exists a normed space $(D, \|\cdot\|_D)$ such that $D(A_t) = D$ for all $t \in \mathbb{R}$ and (the set) $D$ is a dense subspace of $X$.

($A$-iii) $[t \mapsto A_t] \in C(\mathbb{T}, \mathscr{L}(D, X))$.

Furthermore, we assume that the dual family $(A_t^* : X^* \supset D(A_t^*) \to X^*)_{t\in\mathbb{T}}$ fulfills the following conditions.

($A^*$-i) There exists a normed space $(D', \|\cdot\|_{D'})$ such that $D(A_t^*) = D'$ for all $t \in \mathbb{R}$.

($A^*$-iii) $[t \mapsto A_t^*] \in C(\mathbb{T}, \mathscr{L}(D', X^*))$.

*We assume throughout this chapter* that also the conditions

($A$-iv) there exists $\rho \in \mathbb{R}$ such that $\rho + i\mathbb{R} \subset \rho(A_t)$ and
$$\{\, (|\lambda| + 1)(A_t - \lambda)^{-1} : \lambda \in \rho + i\mathbb{R} \,\}$$
is uniformly $(X, X)$-$R$-bounded for all $t \in \mathbb{T}$ and

($A$-v) $D$ is compactly embedded into $X$

hold.

We refer to Section 4.7 for equivalent formulations and to see that the present conditions also imply ($A$-ii) and ($A^*$-ii).

**8.1.2 Theorem**
The set of Floquet exponents $\mathcal{F}set$ is either discrete in $\mathbb{C}\backslash\{0\}$ or $\mathcal{F}set = \mathbb{C}\backslash\{0\}$.

*Proof.*
By Proposition 7.8.1 $\mathcal{F}set = (\mathrm{CS}(\mathfrak{L}'))^{-1}$ and Corollary 7.5.4 yields that $\mathrm{CS}(\mathfrak{L}')$ is an analytic set in $\mathbb{C}\backslash\{0\}$. In particular, $\mathrm{CS}(\mathfrak{L}')$ is either discrete in $\mathbb{C}\backslash\{0\}$ or $\mathrm{CS}(\mathfrak{L}') = \mathbb{C}\backslash\{0\}$. This directly yields the statement. □

**8.1.3 Theorem**
If $\mathcal{F}set$ is discrete then for every $z \in \mathcal{F}set$ there exists a finite set $\{u_z^{(l)}\}_{l\in L_z}$ of Floquet solutions in $\mathcal{F}sol_z$ such that

a) for any at most exponentially increasing solution $u : \mathbb{R} \to X$ of (E) there exist uniquely determined coefficients $\alpha_i \in \mathbb{C}$ where $i \in I := \{\,(z,l) : z \in \mathcal{F}set, l \in L_z\,\}$ with $\alpha_i = 0$ for almost all $i \in I$ such that

$$u = \sum_{(z,l)\in I} \alpha_{(z,l)} u_z^{(l)} \text{ a.e. on } \mathbb{R},$$

b) if $\alpha_i \in \mathbb{C}$ for each $i \in I$ with $\alpha_i = 0$ for almost all $i \in I$ then

$$\sum_{(z,l)\in I} \alpha_{(z,l)} u_z^{(l)}$$

is an at most exponentially increasing solution to (E).

*Remarks on the proof.*
We will prove Theorem 8.1.3 in combination with the next theorem.

**8.1.4 Theorem**
If $\mathcal{F}set = \mathbb{C}\backslash\{0\}$ then there exist

- $n_U \in \mathbb{N}$ and [1] $U^{(l)} : \mathbb{C}\backslash\{0\} \times \mathbb{R} \to X$ for each $l = 1,\dots,n_U$ such that
  - the function $\mathbb{C}\backslash\{0\} \ni z \mapsto U^{(l)}(z,\cdot) \in \mathcal{W}_{0,\mathrm{loc}}(\mathbb{R})$ is analytic for each $l = 1,\dots,n_U$ and
  - for all $z \in \mathbb{C}\backslash\{0\}$ and $l = 1,\dots,n_U$ $[\mathbb{R} \ni t \mapsto U^{(l)}(z,\cdot) \in X] \in \mathcal{F}sol_z$

  and
- a discrete subset $Z \subset \mathbb{C}\backslash\{0\}$ and for each $z \in Z$ a finite set $\{u_z^{(l)}\}_{l\in L_z} \subset \mathcal{F}sol_z$

[1] Using the identification of $\mathbb{C}\backslash\{0\}$ with $\mathbb{R}^2\backslash\{0\}$ we will allow the first component of the argument of $U^{(l)}$ to take values in $\mathbb{R}^2\backslash\{0\}$.

such that

a) for any at most exponentially increasing solution $u : \mathbb{R} \to X$ of (E) there exist for each $l = 1, \ldots, n_U$ $\mu^{(l)} \in C_c^\infty(\mathbb{R}^2\backslash\{0\})$ and coefficients $\alpha_i \in \mathbb{C}$ where $i \in I := \{\,(z,l) : z \in Z, l \in L_z\,\}$ with $\alpha_i = 0$ for almost all $i \in I$ such that[2]
$$u = [t \mapsto \sum_{(z,l)\in I} \alpha_{(z,l)} u_z^{(l)}(t) + \sum_{l=1}^{n_U} \int_{\mathbb{R}^2\backslash\{0\}} U^{(l)}(\xi,t)\mu^{(l)}(\xi)\,d\xi] \text{ a.e. on } \mathbb{R},$$

b) if for each $l = 1, \ldots, n_U$ $\mu^{(l)} \in C_c^\infty(\mathbb{R}^2\backslash\{0\})$ and for each $i \in I$ $\alpha_i \in \mathbb{C}$ with $\alpha_i = 0$ for almost all $i \in I$ then
$$t \mapsto \sum_{(z,l)\in I} \alpha_{(z,l)} u_z^{(l)}(t) + \sum_{l=1}^{n_U} \int_{\mathbb{R}^2\backslash\{0\}} U^{(l)}(\xi,t)\mu^{(l)}(\xi)\,d\xi$$
is an at most exponentially increasing solution to (E).

*Remarks on the proof.*
We mainly follow the proof of [Kuc93] Theorem 3.2.1.

*Proof of Theorem 8.1.3 and Theorem 8.1.4.*

### 8.1.5 Preparation

We remind the reader of the following objects introduced earlier or in the appendix.

$\mathfrak{L}' : \langle\mathfrak{C}_1'\rangle \to \langle\mathfrak{B}_0'\rangle$ is an analytic Fredholm homomorphism between the Banach vector bundles $\langle\mathfrak{C}_1'\rangle$ and $\langle\mathfrak{B}_0'\rangle$, cf. Proposition 7.5.3.

The induced homomorphism on the corresponding spaces of sections is denoted by $\mathfrak{L}'_\Gamma \in \mathscr{L}\big(\Gamma(\mathbb{C}\backslash\{0\}, \langle\mathfrak{C}_1'\rangle), \Gamma(\mathbb{C}\backslash\{0\}, \langle\mathfrak{B}_0'\rangle\big)$, cf. Definition A.7.1.

Furthermore, the corresponding sheaves of germs of sections
$$\mathcal{O}^{\langle\mathfrak{C}_1'\rangle} := \mathcal{O}^{\langle\mathfrak{C}_1'\rangle}(\mathbb{C}\backslash\{0\}) \text{ and } \mathcal{O}^{\langle\mathfrak{B}_0'\rangle} := \mathcal{O}^{\langle\mathfrak{B}_0'\rangle}(\mathbb{C}\backslash\{0\}), \text{ resp.},$$
are BCAF sheaves, cf. Construction B.1.12 and Fact B.1.29.

$\mathfrak{L}'$ induces a homomorphism $\mathfrak{L}'_\mathcal{O} : \mathcal{O}^{\langle\mathfrak{C}_1'\rangle} \to \mathcal{O}^{\langle\mathfrak{B}_0'\rangle}$, cf. Fact B.1.30.

Then $\mathcal{M} := \operatorname{Coker} \mathfrak{L}'_\mathcal{O} := {}^{\mathcal{O}^{\langle\mathfrak{B}_0'\rangle}}\!\big/_{\operatorname{Range} \mathfrak{L}'_\mathcal{O}}$ is coherent, cf. Fact B.1.29.

Furthermore, $\Gamma(\mathbb{C}\backslash\{0\}, \mathcal{M})$ is the quotient module
$$\Gamma(\mathbb{C}\backslash\{0\}, \mathcal{O}^{\langle\mathfrak{B}_0'\rangle})\big/_{\Gamma(\mathbb{C}\backslash\{0\}, \operatorname{Range} \mathfrak{L}'_\mathcal{O})},$$
cf. Definition B.1.21, and we denote by

[2] This representation is, in general, not uniquely determined.

$$\mathcal{P} \in \mathscr{L}(\Gamma(\mathbb{C}\backslash\{0\}, \mathcal{O}^{\langle \mathfrak{B}_0' \rangle}), \Gamma(\mathbb{C}\backslash\{0\}, \mathcal{M}))$$

the natural projection.

We set $\mathcal{Z} := \operatorname{Assoc} \mathcal{M}$. We refer to [Kuc93] Definition 1.5.15 for the definition of the set $\operatorname{Assoc} \mathcal{M}$, but we remark that in particular each $Z \in \mathcal{Z}$ is an irreducible analytic subset of $\mathbb{C}\backslash\{0\}$. Clearly this yields that $Z$ is either of the form $Z = \{z\}$ with some $z \in \mathbb{C}\backslash\{0\}$ or $Z = \mathbb{C}\backslash\{0\}$.

Then, in particular, the sheaf $\mathcal{O}(Z)$ of germs of analytic functions on the analytic set $Z \in \mathcal{Z}$ can be identified with $\mathcal{O}^{\mathbb{C}\backslash\{0\}}$ if $Z = \mathbb{C}\backslash\{0\}$ and with $\mathbb{C}$ if $Z = \{z\}$, cf. [GR65] Section IV.D. In any case, $Z$ consists only of regular points.

Furthermore, from the definition of $\mathcal{Z}$ we obtain $\bigcup_{Z\in\mathcal{Z}} Z = \operatorname{CS}(\mathfrak{L}')$. Thus $\bigcup_{Z\in\mathcal{Z}} Z^{-1} = \mathfrak{Fset}$ since by Proposition 7.8.1 $(\operatorname{CS}(\mathfrak{L}'))^{-1} = \mathfrak{Fset}$.

By [Pal93] Theorem 3.4 for each $Z \in \mathcal{Z}$ there exits a $Z$-Noether operator[3] $\nu_Z : \mathcal{M} \to [\mathcal{O}(Z)]^{n_Z}$ where $\nu_Z = (\nu_Z^{(1)}, \ldots, \nu_Z^{(n_Z)})$.

Then $\nu_Z$ induces a linear map $\hat{\nu}_Z : \Gamma(\mathbb{C}\backslash\{0\}, \mathcal{M}) \to \Gamma(Z, [\mathcal{O}(Z)]^{n_Z})$. In case of $Z = \{z\}$ $\hat{\nu}_Z$ can be identified with the map $\delta_z \hat{\nu}_Z : \Gamma(\mathbb{C}\backslash\{0\}, \mathcal{M}) \to \mathbb{C}^{n_Z}$ and in the case of $Z = \mathbb{C}\backslash\{0\}$ $\hat{\nu}_Z$ is of the form $\hat{\nu}_Z : \Gamma(\mathbb{C}\backslash\{0\}, \mathcal{M}) \to A(\mathbb{C}\backslash\{0\}, \mathbb{C}^{n_Z})$.

By [Kuc93] Theorem 1.7.1 we obtain a characterization of $\operatorname{Coker}(\mathfrak{L}'_\Gamma)$, namely $S'' \in \operatorname{Coker}(\mathfrak{L}'_\Gamma)$ iff there is a finite set $\mathcal{Z}_{\text{fin}} \subset \mathcal{Z}$ and smooth functions $\mu_Z : Z \to \mathbb{C}^n_Z$ such that $S''$ is of the form $S''(s') = \sum_{Z\in\mathcal{Z}_{\text{fin}}} \int_Z \langle \hat{\nu}_Z \mathcal{P} s', \mu_Z \rangle \, dV_Z$ for all $s' \in \Gamma(\mathbb{C}\backslash\{0\}, \langle \mathfrak{B}_0' \rangle)$, where we used the notations of the cited theorem[4]. In particular, $\langle \cdot, \cdot \rangle$ denotes the inner product on $\mathbb{C}^{n_Z}$.

We explain the notation in the given situation. If $Z = \mathbb{C}\backslash\{0\}$ we identify $\mathbb{C}\backslash\{0\}$ with the real manifold $\mathbb{R}^2\backslash\{0\}$ and then $\mu_Z \in C_c^\infty(\mathbb{R}^2\backslash\{0\}, \mathbb{C}^{n_Z}).dV_Z$ denotes the Lebesgue measure $d\lambda$ on $\mathbb{R}^2\backslash\{0\}$. We remark that

$$s' \in \Gamma(\mathbb{C}\backslash\{0\}, \langle \mathfrak{B}_0' \rangle)$$

yields $\mathcal{P}s' \in \Gamma(\mathbb{C}\backslash\{0\}, \mathcal{M})$ and $\hat{\nu}_Z \mathcal{P} s' \in A(\mathbb{C}\backslash\{0\}, \mathbb{C}^{n_Z})$ and thus

$$\langle \nu_Z \mathcal{P} s', \mu_Z \rangle \in C_c^\infty(\mathbb{R}^2\backslash\{0\}, \mathbb{C})$$

is integrable.

[3]For the general definition of Noether operators, we refer to [Pal93] Definition 3.3 and the definition of the sheaf of germs of analytic functions on an analytic set $\mathcal{O}(Z)$ is given, e. g., in [GR65] IV.D.5 Definition.

[4]We remark that of course $\mathcal{Z}_{\text{fin}}$ and $\mu_Z$ are depending on $S''$, although we omit the dependence in the notation.

If $Z$ is of the form $Z = \{z\}$ then $dV_Z$ denotes the point measure at $z$. Thus, after identifying $\mu_Z : \{z\} \to \mathbb{C}^{n_Z}$ with a constant $(\tilde{\mu}_Z^{(1)}, \ldots, \tilde{\mu}_Z^{(n_Z)}) := \delta_z \mu_Z \in \mathbb{C}^{n_Z}$ we obtain $\int_Z \langle \hat{\nu}_Z \mathcal{P} s', \mu_Z \rangle \, dV_Z = \langle \delta_z(\hat{\nu}_Z \mathcal{P} s'), \delta_z \mu_Z \rangle = \sum_{l=1}^{n_Z} \tilde{\mu}_Z^{(l)} \delta_z(\hat{\nu}_Z^{(l)} \mathcal{P} s')$.

We will now prepare the definition of the (fixed) Floquet solutions that occur in the statement of Theorem 8.1.4.

For each $Z \in \mathcal{Z}$, $z \in Z^{-1}$ and $l = 1, \ldots, n_Z$ we define

$$S''^{(l)}_{Z,z} \in (\Gamma(\mathbb{C}\backslash\{0\}, \langle \mathfrak{B}'_0 \rangle))^*$$

by $S''^{(l)}_{Z,z}(s') := \delta_{1/z}(\hat{\nu}_Z^l \mathcal{P} s')$ for all $s' \in \Gamma(\mathbb{C}\backslash\{0\}, \langle \mathfrak{B}'_0 \rangle)$. $S''^{(l)}_{Z,z}$ is well-defined: Indeed, linearity is obvious and continuity follows from [Pal93] Theorem 3.8.

For each $Z \in \mathcal{Z}$, $z \in Z^{-1}$ and $l = 1, \ldots, n_Z$ we obtain $S''^{(l)}_{Z,z} \in \operatorname{Coker} \mathfrak{L}'_\Gamma$: Let $s' \in \Gamma(\mathbb{C}\backslash\{0\}, \langle \mathfrak{C}'_1 \rangle)$. Then for all $z \in \mathbb{C}\backslash\{0\}$ $\gamma_z(\mathfrak{L}'_\Gamma s') \in \operatorname{Range} \mathfrak{L}'_\mathcal{O}$, thus $\mathcal{P}\mathfrak{L}'_\Gamma s' = 0$ in $\Gamma(\mathbb{C}\backslash\{0\}, \mathcal{M})$. Thus indeed $S''^{(l)}_{Z,z}(\mathfrak{L}'_\Gamma s') = 0$.

Therefore by Corollary 7.6.13 for each $Z \in \mathcal{Z}$, $z \in Z^{-1}$ and $l = 1, \ldots, n_Z$ $\mathcal{U}^* S''^{(l)}_{Z,z} \in \operatorname{Coker} \mathcal{L}'_{\Phi'_1}$, i.e. $\mathcal{U}^* S''^{(l)}_{Z,z}$ corresponds to a solution, cf. Proposition 6.5.3.

We will now show that for each $Z \in \mathcal{Z}$, $z \in Z^{-1}$ and $l = 1, \ldots, n_Z$ there exists $d \in \mathbb{N}$ such that if $s' \in \Gamma(\mathbb{C}\backslash\{0\}, \langle \mathfrak{B}'_0 \rangle)$ has a zero[5] of order $d+1$ at $1/z$ then $S''^{(l)}_{Z,z}(s') = 0$: By the definition of the Noether operator $\mu_Z$ we obtain that the induced homomorphism on the stalks $(\nu_Z)_{1/z} : (\mathcal{M})_{1/z} \to ([\mathcal{O}(Z)]^{n_Z})_{1/z}$ is a differential operator between sheaves[6]. We denote by $d$ its order. Now assume that $s'$ indeed has a zero of order $d+1$ at $1/z$, thus $s' = (\cdot - 1/z)^{d+1} g$ for some $g \in \Gamma(\mathbb{C}\backslash\{0\}, \langle \mathfrak{B}'_0 \rangle)$. Then we obtain[7] $(\nu_Z)_{1/z}(\gamma_{1/z}(\mathcal{P}s')) = -\sum_{i=1}^{d+1} (-1)^i \binom{d+1}{i} \gamma_{1/z}((\cdot - 1/z)^i)(\nu_Z)_{1/z}(\gamma_{1/z}(\mathcal{P}(\cdot - 1/z)^{d+1-i} g))$. The evaluation of the germs at $1/z$ of the right hand side yields $\delta_{1/z}((\nu_Z)_{1/z}(\gamma_{1/z}(\mathcal{P}s'))) = 0$, hence $S''^{(l)}_{Z,z}(s') = 0$.

We conclude, that if $s' \in \Gamma(\mathbb{C}\backslash\{0\}, \langle \mathfrak{B}'_0 \rangle)$ is expanded into the power series[8] $\sum_{\alpha=0}^{\infty} s'_\alpha (\cdot - 1/z)^\alpha$ about $1/z$ on some neighborhood $\Omega$ of $1/z$ then $S''^{(l)}_{Z,z}(s') =$

---

[5]By Proposition A.2.8 $\Gamma(\mathbb{C}\backslash\{0\}, \langle \mathfrak{B}'_0 \rangle)$ is isomorphic to $A(\mathbb{C}\backslash\{0\}, \mathcal{W}'_0[0,1])$ and $s' \in \Gamma(\mathbb{C}\backslash\{0\}, \langle \mathfrak{B}'_0 \rangle)$ iff there is $\tilde{s}' \in A(\mathbb{C}\backslash\{0\}, \mathcal{W}'_0[0,1])$ such that $s'(z) = (z, \tilde{s}'(z))$ for all $z \in \mathbb{C}\backslash\{0\}$, cf. the definition of $\mathfrak{b}_0$. We say $s'$ has a zero of order $d+1$ at $1/z$ iff $\tilde{s}'$ has. In that case we write $s'(z) = (z - 1/z)^{d+1} f'(z)$ iff $\tilde{s}'(z) = (z - 1/z)^{d+1} \tilde{f}'(z)$ for some $\tilde{f}' \in A(\mathbb{C}\backslash\{0\}, \mathcal{W}'_0[0,1])$ and $f'(z) = (z, \tilde{f}'(z))$ for all $z \in \mathbb{C}\backslash\{0\}$.

[6]For a definition, we refer to [Pal93] Section 3.

[7]Cf. [Pal68] Proposition 1.1.

[8]The expansion is meant in the sense of footnote 5, i.e. $s'$ is identified with an function in $A(\mathbb{C}\backslash\{0\}, \mathcal{W}'_0[0,1])$.

$S''^{(l)}_{Z,z}(\sum_{\alpha=0}^{d} s'_\alpha(\cdot - 1/z)^\alpha)$. That means that $S''^{(l)}_{Z,z}$ coincides with a functional on $(\mathcal{W}'_0[0,1])^d$. Hence $S''^{(l)}_{Z,z}$ can be represented in the form

$s' \mapsto \sum_{\alpha=0}^{d} \langle f_\alpha, s'_\alpha \rangle_{\mathcal{W}}$

with (a fixed) $f_\alpha \in \mathcal{W}_0[0,1]$ for each $\alpha = 0, \dots, d$. This means that $S''^{(l)}_{Z,z}$ can be represented in the form $s' \mapsto \delta_{1/z}[\Omega \ni \xi \mapsto \sum_{\alpha=0}^{d} \langle f_\alpha, (\partial^\alpha s')(\xi) \rangle_{\mathcal{W}}]$ where by abuse of notation $f_\alpha \in \mathcal{W}_0[0,1]$ for each $\alpha = 0, \dots, d$ may now denote some different but again fixed parameter. Finally, we define $\sigma_\alpha \in \Gamma(\Omega, \langle \mathfrak{B}_0 \rangle)$ to be the constant section $z \mapsto (z, f_\alpha)$ for each $\alpha = 0, \dots, d$ and then we conclude that $S''^{(l)}_{Z,z}$ has the form then $s' \mapsto \sum_{\alpha=0}^{d} \delta_{1/z}\big(\partial^\alpha \langle s', \sigma_\alpha \rangle_\Omega\big)$. Therefore $S''^{(l)}_{Z,z} \in \mathcal{F}func_z$ and by Proposition 7.7.6 $\mathcal{U}^* S''^{(l)}_{Z,z}$ corresponds to a function of Floquet form, i. e. by Proposition 7.7.7 for each $Z \in \mathcal{Z}$, $z \in Z$ and $l = 1, \dots, n_Z$ there exists $u^{(l)}_{Z,z} \in \mathcal{F}sol_z$ such that $\mathrm{F}u^{(l)}_{Z,z} = \mathcal{U}^* S''^{(l)}_{Z,z}$.

We point out that $S''^{(l)}_{Z,z}$ (and thus $u^{(l)}_{Z,z}$) does not depend on $S''$.

Furthermore, in the case of $Z = \mathbb{C}\backslash\{0\}$ the map $\mathbb{C}\backslash\{0\} \ni z \mapsto u^{(l)}_{Z,z}$ is analytic for each $l = 1, \dots, n_Z$: First we note that if we endow $(\Gamma(\mathbb{C}\backslash\{0\}, \langle \mathfrak{B}'_0 \rangle))^*$ with the weak-* topology by Fact 1.5.3 we obtain

$S''^{(l)}_{Z,z} \in A(\mathbb{C}\backslash\{0\}, (\Gamma(\mathbb{C}\backslash\{0\}, \langle \mathfrak{B}'_0 \rangle))^*)$.

Remark 1.5.14 in combination with the proof of Proposition 7.7.6 then yields $[z \mapsto u^{(l)}_{Z,z}] \in A(\mathbb{C}\backslash\{0\}, \mathcal{W}_{0,\mathrm{loc}}(\mathbb{R}))$. △

We are now in the position to prove Theorem 8.1.3 and Theorem 8.1.4.

Let $u$ be an arbitrary at most exponentially increasing solution of (E).

By Proposition 6.5.3 $\mathrm{F}u \in \operatorname{Coker} \mathcal{L}'_{\Phi'_1}$ and by Corollary 7.6.13 we obtain $S'' := (\mathcal{U}^*)^{-1}(\mathrm{F}u) \in \operatorname{Coker} \mathfrak{L}'_\Gamma$.

As in Preparation 8.1.5 we denote by $\mu_Z$ and $\mathcal{Z}_{\mathrm{fin}}$ the corresponding objects that are used to represent $S''$ in [Kuc93] Theorem 1.7.1.

Then for all $\phi' \in \Phi'_0$ and $s' := \mathcal{U}\phi' \in \Gamma(\mathbb{C}\backslash\{0\}, \langle \mathfrak{B}'_0 \rangle)$ we obtain $(\mathrm{F}u)(\phi') = (\mathcal{U}^*)^{-1}(\mathrm{F}u)(\mathcal{U}\phi') = S''(s') = \sum_{Z \in \mathcal{Z}_{\mathrm{fin}}} \int_Z \langle \hat{\nu}_Z \mathcal{P} s', \mu_Z \rangle \, dV_Z$.

We first treat the case that $\mathcal{F}set$ is discrete. Then each $Z \in \mathcal{Z}$ is of the form $Z = \{z\}$.

We write $u^{(l)}_z$ instead of $u^{(l)}_{Z,z}$ for each $l = 1, \dots, n_Z$ if $Z \in \mathcal{Z}$ and $\{z\} = Z^{-1}$.

Every summand in the above sum is given by

$$\int_{\{z\}} \langle \hat{\nu}_{\{z\}} \mathcal{P} s', \mu_{\{z\}} \rangle \, dV_{\{z\}} = \sum_{l=1}^{n_{\{z\}}} \tilde{\mu}_{\{z\}}^{(l)} u_{1/z}^{(l)}(s')$$

for the corresponding $z \in \{z\} = Z$.

Therefore

$$(\mathrm{F}u)(\phi') =$$

$$\sum_{\{z\} \in \mathcal{Z}_{\mathrm{fin}}} \sum_{l=1}^{n_{\{z\}}} \tilde{\mu}_{\{z\}}^{(l)} u_{1/z}^{(l)}(s') =$$

$$\sum_{\{z\} \in \mathcal{Z}_{\mathrm{fin}}} \sum_{l=1}^{n_{\{z\}}} \tilde{\mu}_{\{z\}}^{(l)} (\mathcal{U}^* u_{1/z}^{(l)})(\phi') =$$

$$\sum_{\{z\} \in \mathcal{Z}_{\mathrm{fin}}} \sum_{l=1}^{n_{\{z\}}} \tilde{\mu}_{\{z\}}^{(l)} (\mathrm{F} u_{1/z}^{(l)})(\phi') =$$

$$\Big(\mathrm{F}\big(\sum_{\{z\} \in \mathcal{Z}_{\mathrm{fin}}} \sum_{l=1}^{n_{\{z\}}} \tilde{\mu}_{\{z\}}^{(l)} u_{1/z}^{(l)}\big)\Big)(\phi').$$

We conclude that $u = \sum_{\{z\} \in \mathcal{Z}_{\mathrm{fin}}} \sum_{l=1}^{n_{\{z\}}} \tilde{\mu}_{\{z\}}^{(l)} u_{1/z}^{(l)}$ is indeed of the stated form.

We remark that every discrete set in $\mathbb{C}\backslash\{0\}$ is countable, thus $\{ u_{1/z}^{(l)} : \{1/z\} \in \mathcal{Z}_{\mathrm{fin}}{}_{l=1}^{n_{\{1/z\}}} \}$ is countable. Thus uniqueness of the coefficients can be obtained by simply omitting solutions $u_{1/z}^{(l)}$ that would allow different representations.

In the case that $\mathcal{F}set = \mathbb{C}\backslash\{0\}$, additionally the summand

$$\int_{\mathbb{R}^2\backslash\{0\}} \langle \hat{\nu}_{\mathbb{C}\backslash\{0\}} \mathcal{P} s', \mu_{\mathbb{C}\backslash\{0\}} \rangle \, d\lambda$$

occurs.

We set $\nu := \nu_{\mathbb{C}\backslash\{0\}}$, $n := n_{\mathbb{C}\backslash\{0\}}$, $\mu := \mu_{\mathbb{C}\backslash\{0\}}$, $S_z''^{(l)} := S''_{\mathbb{C}\backslash\{0\},z}{}^{(l)}$ and $U_z^{(l)} := u_{\mathbb{C}\backslash\{0\},1/z}^{(l)}$ for all $l = 1, \ldots, n$ and $z \in \mathbb{C}\backslash\{0\}$. Then we obtain

$$\int_{\mathbb{R}^2\backslash\{0\}} \langle \hat{\nu} \mathcal{P} s', \mu \rangle \, d\lambda =$$

$$\int_{\mathbb{R}^2\backslash\{0\}} \sum_{l=1}^{n} (\hat{\nu} \mathcal{P} s')_l(z) \mu^{(l)}(z) \, dz =$$

$$\sum_{l=1}^{n} \int_{\mathbb{R}^2\backslash\{0\}} \delta_z(\hat{\nu}^{(l)} \mathcal{P} s') \mu^{(l)}(z) \, dz =$$

$$\sum_{l=1}^{n}\int_{\mathbb{R}^2\setminus\{0\}} {S''_{1/z}}^{(l)}(s')\mu^{(l)}(z)\,dz =$$

$$\sum_{l=1}^{n}\int_{\mathbb{R}^2\setminus\{0\}} \mathcal{U}^*{S''_{1/z}}^{(l)}(\mathcal{U}^{-1}s')\mu^{(l)}(z)\,dz =$$

$$\sum_{l=1}^{n}\int_{\mathbb{R}^2\setminus\{0\}} (\mathrm{F}U_z^{(l)})(\phi')\mu^{(l)}(z)\,dz =$$

$$\sum_{l=1}^{n}\int_{\mathbb{R}^2\setminus\{0\}}\int_{\mathbb{R}} \langle\phi'(t), U_z^{(l)}(t)\rangle_X\,dt\mu^{(l)}(z)\,dz =$$

$$\int_{\mathbb{R}}\sum_{l=1}^{n}\int_{\mathbb{R}^2\setminus\{0\}} \langle\phi'(t), U_z^{(l)}(t)\rangle_X\mu^{(l)}(z)\,dz\,dt =$$

$$\int_{\mathbb{R}}\sum_{l=1}^{n}\langle\phi'(t), (\int_{\mathbb{R}^2\setminus\{0\}} U_z^{(l)}(\cdot)\mu^{(l)}(z)\,dz)(t)\rangle_X\,dt =$$

$$(\mathrm{F}(\sum_{l=1}^{n}\int_{\mathbb{R}^2\setminus\{0\}} U_z^{(l)}(\cdot)\mu^{(l)}(z)\,dz))(\phi').$$

This finishes the proof. □

**8.1.6 Theorem**

The following are equivalent.

(1) $\mathcal{F}set = \mathbb{C}\setminus\{0\}$.

(2) For each $a > 0$ there is a solution $u \neq 0$ and $c > 0$ such that
$\|u\|_{\mathcal{W}_1[k,k+1]} \leq c\exp(-a|k|)$ for all $k \in \mathbb{Z}$.

(3) There is a solution $u \neq 0$ and $a > 0$, $c > 0$ such that
$\|u\|_{\mathcal{W}_1[k,k+1]} \leq c\exp(-a|k|)$ for all $k \in \mathbb{Z}$.

*Proof.*

"(1)⇒(2)":

As a first step, we will show that there exists $0 \neq S \in \Gamma(\mathbb{C}\setminus\{0\}, \langle\mathfrak{C}_1\rangle)$ such that $\mathfrak{L}_\Gamma S = 0$.

Proposition 7.8.1 in combination with the assumption (1) yields $\mathrm{S}(\mathfrak{L}) = \mathcal{F}set = \mathbb{C}\setminus\{0\}$, hence there is $z \in \mathbb{C}\setminus\{0\}$ such that

$$\dim\operatorname{Ker}\mathfrak{L}_z = \min\{\dim\operatorname{Ker}\mathfrak{L}_\xi : \xi \in \mathbb{C}\setminus\{0\}\} \geq 1.$$

For each $\xi \in \mathbb{B}_z$ we denote by $L(\xi)$ the trivialized induced map $\mathcal{W}_1(\mathbb{T}) \xrightarrow{((\mathfrak{c}_1{}^{(z)})_\xi)^{-1}} (\mathfrak{C}_1)_\xi \xrightarrow{\mathfrak{L}_\xi} (\mathfrak{B}_0)_\xi \xrightarrow{(\mathfrak{b}_0)_\xi} \mathcal{W}_0[0,1]$ of the bundle homomorphism $\mathfrak{L}$. Thus by Proposition 7.5.1 $[\xi \mapsto L(\xi)] \in A(\mathbb{B}_z, \mathscr{L}(\mathcal{W}_1(\mathbb{T}), \mathcal{W}_0[0,1])$ is Fredholm operator-valued. Hence by [Kuc93] Corollary 1.2.14 the map $\xi \mapsto \dim \operatorname{Ker} L(\xi)$ is upper semi-continuous. In particular, there exists a neighborhood $\Omega \overset{\circ}{\subset} \mathbb{B}_z$ of $z$ such that $\xi \mapsto \dim \operatorname{Ker} L(\xi)$ is constant on $\Omega$. We conclude that $\xi \mapsto \dim \operatorname{Ker} \mathfrak{L}_\xi$ is constant on $\Omega$.

We endow $\mathfrak{K} := \bigcup_{\xi \in \Omega}\{\xi\} \times \operatorname{Ker} \mathfrak{L}_\xi \subset \mathfrak{C}_{1|\Omega}$ with the induced topology. Furthermore, for all $\xi \in \Omega$ we understand $\mathfrak{K}_\xi := \{\xi\} \times \operatorname{Ker} \mathfrak{L}_\xi$ as a (closed) linear subspace of $(\mathfrak{C}_1)_\xi$. Then $\nu := [\mathfrak{K} \ni (\xi, f) \mapsto \xi \in \Omega]$ is a bundle projection. By [Kuc93] Theorem 1.6.13 there exists a trivialization for $\nu$ such that its equivalence class $\langle \mathfrak{K} \overset{\nu}{\succ} \Omega \rangle$ is a subbundle of $\langle \mathfrak{C}_1 \rangle$.

By choosing $\Omega$ smaller, if need be, by Proposition A.2.8 and by Proposition A.1.11 we can assume w.l.o.g. that $\Gamma(\Omega, \langle \mathfrak{K} \overset{\nu}{\succ} \Omega \rangle)$ is isomorphic to $A(\Omega, \operatorname{Ker} \mathfrak{L}_z)$. Hence there is $s \in \Gamma(\Omega, \langle \mathfrak{K} \overset{\nu}{\succ} \Omega \rangle)$ such that $s(z) \neq 0$. By Remark A.6.5 and Remark A.3.2 $s \in \Gamma(\Omega, \langle \mathfrak{C}_1 \rangle)$. Furthermore, $\mathfrak{L}_\xi(s(\xi)) = 0$ for all $\xi \in \Omega$. Thus[9] $\gamma_z(s) \in \operatorname{Ker} \mathfrak{L}_{\mathcal{O}}$. By [Kuc93] Theorem 1.5.9 ii) in combination with Fact B.1.30 there is $\sigma \in \Gamma(\mathbb{C}\backslash\{0\}, \operatorname{Ker} \mathfrak{L}_{\mathcal{O}})$ such that $\sigma(z) = \gamma_z(s)$. By Definition B.1.20 $\sigma \in \Gamma(\mathbb{C}\backslash\{0\}, \mathcal{O}^{\langle \mathfrak{C}_1 \rangle}(\mathbb{C}\backslash\{0\}))$ and thus by Remark B.1.14 there is $S \in \Gamma(\mathbb{C}\backslash\{0\}, \langle \mathfrak{C}_1 \rangle)$ such that $\sigma(\xi) = \gamma_\xi(S)$ for all $\xi \in \mathbb{C}\backslash\{0\}$. Finally, $S(z) = (\sigma(z))(z) = s(z) \neq 0$ and for each $\xi \in \mathbb{C}\backslash\{0\}$ $\gamma_\xi(\mathfrak{L}_\Gamma S) = \mathfrak{L}_{\mathcal{O}}(\sigma(\xi)) = 0$, hence $\mathfrak{L}_\Gamma S = 0$.

Now, Construction 7.6.6 and Corollary 7.6.7 yield $0 \neq u := \mathcal{U}^{-1} S \in \Phi_1$ and $\mathcal{L}_{\mathcal{W}_{1,\mathrm{loc}}(\mathbb{R})} u = \mathcal{L}_{\Phi_1} u = (\mathcal{U}^{-1} \circ \mathfrak{L}_\Gamma \circ \mathcal{U}) u = (\mathcal{U}^{-1} \circ \mathfrak{L}_\Gamma) S = 0$, hence by Proposition 4.5.3 $u$ is a solution. This yields (2).

"(2)⇒(3)": This is obvious.

"(3)⇒(1)": Let $\|u\|_{\mathcal{W}_1[k,k+1]} \leq c\exp(-a|k|)$, $a > 0$ and $c > 0$ such that $\|u\|_{\mathcal{W}_1[k,k+1]} \leq c\exp(-a|k|)$ for all $k \in \mathbb{Z}$. Then $0 \neq u \in \Phi_{1,a}$ and hence Construction 7.6.6 yields $0 \neq \mathcal{U}u \in \Gamma(\mathbb{A}_\alpha, \langle \mathfrak{C}_1 \rangle)$. In combination with Proposition A.2.8 we obtain that there is $z \in \mathbb{A}_\alpha$ and am open neighborhood $z \in \Omega \overset{\circ}{\subset} \mathbb{A}_\alpha$ such that $(\mathcal{U}u)(\xi) \neq 0$ for all $\xi \in \Omega$. By Proposition 4.5.3 $\mathcal{L}_{\mathcal{W}_{1,\mathrm{loc}}(\mathbb{R})} u = 0$ and then by Corollary 7.6.7 $\mathfrak{L}_{\Gamma|\mathbb{A}_a}(\mathcal{U}u) = 0$ holds. For each $\xi \in \Omega$ we obtain $\mathfrak{L}_\xi\big((\mathcal{U}u)(\xi)\big) = 0$, thus $0 \neq (\mathcal{U}u)(\xi) \in \operatorname{Ker} \mathfrak{L}_\xi$. Hence $\xi \in \mathrm{S}(\mathfrak{L})$. Therefore, in combination with by Proposition 7.8.1 $\Omega \subset \mathrm{S}(\mathfrak{L}) = \mathcal{F}set$ and thus $\mathcal{F}set$ is not

[9]We remind the reader that $\mathfrak{L}_{\mathcal{O}}$ denotes the induced sheaf homomorphism by $\mathfrak{L}$, cf. Fact B.1.30 and $\operatorname{Ker} \mathfrak{L}_{\mathcal{O}}$ is a subsheaf of $\mathcal{O}^{\langle \mathfrak{C}_1 \rangle}(\mathbb{C}\backslash\{0\})$, cf. Definition B.1.23.

discrete. Then Theorem 8.1.2 yields $\mathcal{F}set = \mathbb{C}\backslash\{0\}$. □

We directly obtain:

**8.1.7 Corollary**
If $\mathcal{F}set = \mathbb{C}\backslash\{0\}$ then there is a solution $0 \neq u \in L_p(\mathbb{R}, X)$.

### Bloch Property

We finish this section by analyzing the so-called Bloch property. First we remark that the following direct consequence of the first three theorems in this chapter holds.

**8.1.8 Corollary**
If there exists a non-vanishing, bounded[10] solution then there exists at least one Bloch solution.

*Proof.*
If there is a bounded solution $u \neq 0$ to (E) then by Theorem 8.1.2 we are either in the situation of Theorem 8.1.3 or Theorem 8.1.4. In any case we particularly obtain $\mathcal{F}set \neq \emptyset$ since there must exist at least one Floquet solution to represent $u$. Then Proposition 7.8.1 yields $\mathcal{B}set = \mathcal{F}set \neq \emptyset$, in other words there exists a Bloch solution. □

We now show that under an additional assumption the Bloch property holds.

**8.1.9 Theorem** (*Bloch Property*)
Assume that the bundle $\langle \mathfrak{C}_1' \rangle$ is trivial.
Furthermore, assume that there exists an at most exponentially increasing solution $u \neq 0$ with corresponding constants $c, a > 0$, i. e. $\|u\|_{\mathcal{W}_0[k,k+1]} \leq c\exp(a|k|)$ for all $k \in \mathbb{Z}$.
Then there is $z \in \mathcal{B}set$ such that $\exp(-a) \leq |z| \leq \exp(a)$.
(For conditions when $\langle \mathfrak{C}_1' \rangle$ is trivial we refer to Remark 8.1.11.)

*Proof.*
We will prove the statement by contradiction. Assume that there is no $z \in \mathcal{B}set$ with $\exp(-a) \leq |z| \leq \exp(a)$. Then in particular by Theorem 8.1.2 and Proposition 7.8.1 $\mathcal{B}set = \mathcal{F}set$ is discrete in $\mathbb{C}\backslash\{0\}$ and we conclude that there exists $\epsilon > 0$ such that for all $z \in \mathcal{B}set$ $|z| < \exp(-(a+\epsilon))$ or $|z| > \exp(a+\epsilon)$.

---
[10](a. e. on $\mathbb{R}$)

We will now show that $(\mathrm{F}u)(\phi') = 0$ for all $\phi' \in \Phi'_{0,a+\epsilon}$. Then, since $\Phi'_0 \subset \Phi'_{0,a+\epsilon}$, by Proposition 6.5.2 we obtain the contradiction $u = 0$.

It suffices to show that $\mathcal{L}'_{\Phi'_{1,a+\epsilon}} : \Phi'_{1,a+\epsilon} \to \Phi'_{0,a+\epsilon}$ is surjective: Indeed then for each $\phi' \in \Phi'_{0,a+\epsilon}$ there is $\psi' \in \Phi'_{1,a+\epsilon}$ with $\mathcal{L}'_{\Phi'_{1,a+\epsilon}}\psi' = \phi'$ and $(\mathrm{F}u)(\phi') = (\mathrm{F}u)(\mathcal{L}'_{\Phi'_{1,a+\epsilon}}\psi') = \langle \mathcal{L}'_{\mathcal{W}'_{1,\mathrm{loc}}(\mathbb{R})}\psi', u\rangle = 0$ by the definition of solutions.

By Corollary 7.6.12 surjectivity of $\mathcal{L}'_{\Phi'_{1,a+\epsilon}}$ is equivalent to surjectivity of $\mathfrak{L}'_{\Gamma|\mathbb{A}_{a+\epsilon}}$.

By Proposition 7.8.1 $\mathscr{B}set = (\mathrm{CS}(\mathfrak{L}'))^{-1}$ and hence for all $z \in \mathrm{CS}(\mathfrak{L}')$ $|z| > \exp(a+\epsilon)$ or $|z| < exp(-(a+\epsilon))$. In particular, $\mathrm{CS}(\mathfrak{L}') \cap \mathbb{A}_{a+\epsilon} = \emptyset$. By Proposition A.5.2 in combination with Proposition A.1.10 and Remark 7.4.4 there is a trivializing map $\mathfrak{c}' : (\mathfrak{C}'_1)_{|\mathbb{A}_{a+\epsilon}} \to \mathbb{A}_{a+\epsilon} \times \mathcal{W}'_0[0,1]$. The same argument[11] as in the step "$\mathscr{F}set \subset (\mathrm{CS}(\mathfrak{L}'))^{-1}$" of the proof of Proposition 7.8.1 now yields that $\mathfrak{L}'_{\Gamma|\mathbb{A}_{a+\epsilon}}$ is surjective. This finishes the proof. □

**8.1.10 Corollary** (*Bloch Property, Classic Version*)
Assume that the bundle $\langle\mathfrak{C}'_1\rangle$ is trivial.
If there exists a non-vanishing, bounded[12] solution then there exists at least one bounded[12] Bloch solution.

*Proof.*
We recall that by Proposition 7.8.1 $\mathscr{B}set = \mathscr{F}set$. If $\mathscr{B}set = \mathbb{C}\backslash\{0\}$ then the statement directly follows. Thus by Theorem 8.1.2 we can assume that $\mathscr{B}set$ is discrete in $\mathbb{C}\backslash\{0\}$. If we denote the bounded solution by $u$ then for all $a > 0$ there is $c > 0$ such that $\|u\|_{\mathcal{W}_0[k,k+1]} \leq c\exp(a|k|)$ for all $k \in \mathbb{Z}$. Thus by Theorem 8.1.9 there exists $(z_n)_{n\in\mathbb{N}} \in \mathscr{B}set$ such that $\exp(-1/n) \leq |z_n| \leq \exp(1/n)$ for all $n \in \mathbb{N}$ and hence $|z_n| \overset{n\to\infty}{\longrightarrow} 1$. Discreteness of $\mathscr{B}set$ in combination with compactness of $\{\, z \in \mathbb{C} : |z| = 1 \,\}$ yields that there is $z_0 \in \mathscr{B}set$ such that $|z_0| = 1$. There is $u_0 \in \mathscr{B}sol_{z_0}$ and thus $\|u_0(t)\|_X = \|g(t)\|_X$ a. e. on $\mathbb{R}$ for some $g \in L_p(\mathbb{T}, X)$. By Remark 6.2.4 $g$ has a continuous representant and this directly yields the assertion. □

**8.1.11 Remark**
We remark that by Remark 7.4.4 $\langle\mathfrak{C}'_1\rangle$ is a bundle with fiber $\mathcal{W}'_0[0,1]$. Then by [Bun68] § 8 a sufficient condition for the bundle $\langle\mathfrak{C}'_1\rangle$ to be trivial is that the so-called structure group $\mathscr{L}(\mathcal{W}'_0[0,1])$ is contractible.

---

[11]We remark—w. r. t. footnote 8 in said proof (on page 74)—that if need be we might substitute $\epsilon$ by $\epsilon/2$.

[12](a. e. on $\mathbb{R}$)

In particular, if $X$ is a ($\mathbb{C}$-valued) $L_p(\Omega)$-space then

$$\mathcal{W}_0'[0,1] = L_p([0,1], L_p(\Omega)) = L_p([0,1] \times \Omega)$$

is also a $L_p$-space. Hence by [Mit70] Proposition 5 $\mathscr{L}(\mathcal{W}_0'[0,1])$ is contractible and $\langle \mathfrak{C}_1' \rangle$ is trivial.

# Appendix A

# Analytic Banach Vector Bundles

Throughout this chapter let $\varOmega \overset{\circ}{\subset} \mathbb{C}$.

## A.1 Analytic Banach Vector Bundles

**A.1.1 Definition** (*Analytic Banach Vector Bundle*)
Let $\mathcal{E}$ be a topological space and $p : \mathcal{E} \to \varOmega$ a surjective, continuous function such that for each $x \in \varOmega$ the so-called *fiber* $\mathcal{E}_x := p^{-1}(\{x\})$ has a Banach space structure, whose topology (that comes from the norm) coincides with the topology induced from $\mathcal{E}$. $\varOmega$ is called *base space*, $\mathcal{E}$ is called *total space* and $p$ is called *(bundle) projection.*

For each $U \overset{\circ}{\subset} \varOmega$ we set $\mathcal{E}_{|U} := p^{-1}(U)$.

Let $\{U_\lambda\}_{\lambda \in \varLambda}$ be an open cover of $\varOmega$. Furthermore, suppose that for each $\lambda \in \varLambda$ there exists a Banach space $B_\lambda$ and a homeomorphism $\phi^{(\lambda)} : \mathcal{E}_{|U_\lambda} \to U_\lambda \times B_\lambda$. Then $\{\phi^{(\lambda)}\}_{\lambda \in \varLambda}$ is called a *trivialization (for p)* if for all $\lambda \in \varLambda$

(a) the diagram

$$\begin{array}{ccc} \mathcal{E}_{|U_\lambda} & \xrightarrow{\phi^{(\lambda)}} & U_\lambda \times B_\lambda \\ & {\scriptstyle p}\searrow \quad \swarrow{\scriptstyle \nu} & \\ & U_\lambda & \end{array}$$

(where $\nu$ denotes the *natural projection* $\nu : U_\lambda \times B_\lambda \to U_\lambda$ given by $\nu(u, b) := u$) commutes, i. e. $p = \nu \circ \phi^{(\lambda)}$ and

(b) $\phi^{(\lambda)}$ induces on each fiber $\mathcal{E}_x$ (where $x \in U_\lambda$) an isomorphism $\phi_x^{(\lambda)} : \mathcal{E}_x \to$

$B_\lambda$, i. e. the map

$$\phi_x^{(\lambda)} : \mathcal{E}_x \overset{(\phi^{(\lambda)})_{|\mathcal{E}_x}}{\longrightarrow} \{x\} \times B_\lambda \overset{\cong}{\longrightarrow} B_\lambda$$

(where $\cong$ denotes the *natural identification* given by $\{x\} \times B_\lambda \ni (x,b) \mapsto b \in B_\lambda$) is an (Banach space) isomorphism and

(c) for all $\kappa \in \Lambda$ with $U_\lambda \cap U_\kappa \neq \emptyset$ the so-called *transition function (from $B_\lambda$ to $B_\kappa$)*

$$\Phi_x^{(\lambda,\kappa)} : B_\lambda \overset{\left(\phi_x^{(\lambda)}\right)^{-1}}{\longrightarrow} \mathcal{E}_x \overset{\phi_x^{(\kappa)}}{\longrightarrow} B_\kappa$$

is analytically depending on $x \in U_\lambda \cap U_\kappa \neq \emptyset$, i. e. $[x \mapsto \Phi_x^{(\lambda,\kappa)}] \in A(U_\lambda \cap U_\kappa, \mathscr{L}(B_\lambda, B_\kappa))$.

In this case, $\{U_\lambda\}_{\lambda\in\Lambda}$ is called the *associated trivializing cover.*

Two trivializations for $p$ are called *equivalent* if their union satisfies (c).

A *nonempty* equivalence class of this relation is called an *(analytic Banach vector) bundle (over $\Omega$)* and is denoted by $\langle \mathcal{E} \overset{p}{\succ} \Omega\rangle$ for short. If $\{\phi^{(\lambda)}\}_{\lambda\in\Lambda}$ is a representant of $\langle \mathcal{E} \overset{p}{\succ} \Omega\rangle$, then each $\phi^{(\lambda)}$ is called a *trivializing map (for $\langle \mathcal{E} \overset{p}{\succ} \Omega\rangle$).*

**A.1.2 Remark**

In the situation of Definition A.1.1 the transition function $\Phi_x^{(\kappa,\lambda)}$ from $B_\kappa$ to $B_\lambda$ (where $U_\lambda \cap U_\kappa \neq \emptyset$) obviously coincides with $\left(\Phi_x^{(\lambda,\kappa)}\right)^{-1}$. Thus $[x \mapsto \Phi_x^{(\lambda,\kappa)}] \in A(U_\lambda \cap U_\kappa, \mathscr{L}(B_\lambda, B_\kappa))$ iff $[x \mapsto \Phi_x^{(\kappa,\lambda)}] \in A(U_\kappa \cap U_\lambda, \mathscr{L}(B_\kappa, B_\lambda))$ (cf. [Cha85] Theorems 7.17, 5.9 and 14.13).

**A.1.3 Remark**

The notion of equivalence introduced in Definition A.1.1 is indeed an equivalence relation: Reflexivity and symmetry are obvious. In order to show transitivity, let $\{\phi^{(\lambda)}\}_{\lambda\in\Lambda}$, $\{\psi^{(\kappa)}\}_{\kappa\in K}$ and $\{\tilde{\phi}^{(\tilde{\lambda})}\}_{\tilde{\lambda}\in\tilde{\Lambda}}$ be trivializations for a bundle projection $p : \mathcal{E} \to \Omega$ such that $\{\phi^{(\lambda)}\}_{\lambda\in\Lambda} \sim \{\psi^{(\lambda)}\}_{\lambda\in\Lambda}$ and $\{\psi^{(\lambda)}\}_{\lambda\in\Lambda} \sim \{\tilde{\phi}^{(\tilde{\lambda})}\}_{\tilde{\lambda}\in\tilde{\Lambda}}$ where $\sim$ denotes equivalence. Let $[\phi : \mathcal{E}_{|U} \to U \times B] \in \{\phi^{(\lambda)}\}_{\lambda\in\Lambda}$ and $[\tilde{\phi} : \mathcal{E}_{|\tilde{U}} \to \tilde{U} \times \tilde{B}] \in \{\tilde{\phi}^{(\tilde{\lambda})}\}_{\tilde{\lambda}\in\tilde{\Lambda}}$ such that $U \cap \tilde{U} \neq \emptyset$ and for each $x \in U \cap \tilde{U}$ we denote the transition function $B \overset{(\phi_x)^{-1}}{\longrightarrow} \mathcal{E}_x \overset{\tilde{\phi}_x}{\longrightarrow} \tilde{B}$ by $\Phi_x$. Let $\xi \in U \cap \tilde{U}$. We will show that there is $O \overset{\circ}{\subset} U \cap \tilde{U}$ with $\xi \in O$ such that $[x \mapsto \Phi_x] \in A(O, \mathscr{L}(B, \tilde{B}))$. Then in combination with Remark A.1.2, we obtain $\{\phi^{(\lambda)}\}_{\lambda\in\Lambda} \sim \{\tilde{\phi}^{(\tilde{\lambda})}\}_{\tilde{\lambda}\in\tilde{\Lambda}}$. Let $[\psi : \mathcal{E}_{|V} \to V \times C] \in \{\psi^{(\kappa)}\}_{\kappa\in K}$ such that $\xi \in V$. Then $O := V \cap U \cap \tilde{U} \overset{\circ}{\subset} U \cap \tilde{U}$ and $\xi \in O$. For each $x \in O$ $\Phi_x$ can be written as $B \xrightarrow{(\phi_x)^{-1}} \mathcal{E}_x \xrightarrow{\psi_x} C \xrightarrow{(\psi_x)^{-1}} \mathcal{E}_x \xrightarrow{\tilde{\phi}_x} \tilde{B}$ (with $\mathcal{E}_x \to C \to \mathcal{E}_x$ equal to $\mathrm{Id}$) as a "product" of $[x \mapsto \psi_x \circ (\phi_x)^{-1}] \in A(O, \mathscr{L}(B, C))$ and $[x \mapsto \tilde{\phi}_x \circ (\psi_x)^{-1}] \in A(O, \mathscr{L}(C, \tilde{B}))$ and thus by Fact 1.5.10 $[x \mapsto \Phi_x] \in A(O, \mathscr{L}(B, \tilde{B}))$.

### A.1.4 Remark

In the situation of Definition A.1.1 condition (b) is equivalent to the condition

(b') $\phi^{(\lambda)}$ induces on each fiber $\mathcal{E}_x$ ($x \in U_\lambda$) a linear map $\phi_x^{(\lambda)} : \mathcal{E}_x \to B_\lambda$, i. e. the map

$$\phi_x^{(\lambda)} : \mathcal{E}_x \xrightarrow{(\phi^{(\lambda)})_{|\mathcal{E}_x}} \{x\} \times B_\lambda \xrightarrow{\cong} B_\lambda$$

(where $\cong$ again denotes the natural identification) is linear

since $\phi$ is homeomorphic and thus the restrictions $(\phi^{(\lambda)})_{|\mathcal{E}_x} : \mathcal{E}_x \to \{x\} \times B_\lambda$ are continuous and bijective.

### A.1.5 Proposition (*On the Notation* $\langle \mathcal{E} \overset{p}{\succ} \Omega \rangle$)

Let $\{E_\lambda\}_{\lambda\in\Lambda}$, $\{U_\lambda\}_{\lambda\in\Lambda}$ be families of topological spaces, $\{B_\lambda\}_{\lambda\in\Lambda}$ a family of Banach spaces and $\{\phi^{(\lambda)} : E_\lambda \to U_\lambda \times B_\lambda\}_{\lambda\in\Lambda}$ a family of homeomorphisms such that $\{\phi^{(\lambda)}\}_{\lambda\in\Lambda}$ is a representant of a bundle $\langle \mathcal{E} \overset{p}{\succ} \Omega \rangle$.
Then the topological spaces $\mathcal{E}$ and $\Omega$ and the map $p : \mathcal{E} \to \Omega$ can be recovered, i. e. they are uniquely determined. Furthermore, for each $x \in \Omega$, the Banach space structure of $\mathcal{E}_x$ can be recovered up to equivalence of the norm.

*Proof.*
Let $\mathcal{E}$, $\Omega$, $p$ and $\{\mathcal{E}_x\}_{x\in\Omega}$ be as in Definition A.1.1 (such that $\{\phi^{(\lambda)}\}_{\lambda\in\Lambda}$ is a trivialization for $p$). The topological space $\Omega$ is determined by $\Omega = \bigcup_{\lambda\in\Lambda} U_\lambda$ and the neighborhood basises of $x \in U_\lambda$ (which are also neighborhood basises in $\Omega$ since $U_\lambda$ is open in $\Omega$) (see [Que01] 2.8, [Que01] 2.9 and [Que01] 2.10). Analogously, the topological space $\mathcal{E}$ is determined by $\mathcal{E} = \bigcup_{\lambda\in\Lambda} E_\lambda$ and the neighborhood basises of $e \in E_\lambda$ (which are also neighborhood basises in $\mathcal{E}$ since $E_\lambda = \mathcal{E}_{|U_\lambda} = p^{-1}(U)$ is open in $\mathcal{E}$). Furthermore, for each $e \in \mathcal{E}$ there exists $\lambda \in \Lambda$ such that $e \in \mathcal{E}_{|U_\lambda}$ and thus $p(e)$ is determined by the first coordinate of $\phi^{(\lambda)}(e)$. Finally, for each $x \in \Omega$ scalar multiplication, addition and the zero vector in $\mathcal{E}_x$ is uniquely determined by $\alpha e = (\phi_x^{(\lambda)})^{-1}(\alpha \phi_x^{(\lambda)} e)$, $e + \tilde{e} =$

$\left(\phi_x^{(\lambda)}\right)^{-1}\left(\phi_x^{(\lambda)}e+\phi_x^{(\lambda)}\tilde{e}\right)$ and $0=\left(\phi_x^{(\lambda)}\right)^{-1}(0)$, resp. for $\alpha \in \mathbb{C}$, $e, \tilde{e} \in \mathcal{E}_x$ and $\lambda \in \Lambda$ such that $x \in U_\lambda$. A complete norm on $\mathcal{E}_x$ is given by $\|e\|_{\mathcal{E}_x} := \|\phi_x^{(\lambda)} e\|_{B_\lambda}$. If $\|\cdot\|$ is another complete norm on $\mathcal{E}_x$ then $\|e\| \leq \|\left(\phi_x^{(\lambda)}\right)^{-1}\|_{\mathscr{L}(B_\lambda,(\mathcal{E}_x,\|\cdot\|))}\|\phi_x^{(\lambda)}e\|_{B_\lambda} = \|\left(\phi_x^{(\lambda)}\right)^{-1}\|_{\mathscr{L}(B_\lambda,(\mathcal{E}_x,\|\cdot\|))}\|e\|_{\mathcal{E}_x}$ and thus by [Wer05] Korollar IV.3.5 $\|\cdot\|_{\mathcal{E}_x}$ and $\|\cdot\|$ are equivalent. □

**A.1.6 Remark**
We will see in Example A.4.8 that the "nonuniqueness" of the Banach space norms that occurred in Proposition A.1.5 is in some sense compatible with the structure of bundles.

**A.1.7 Remark**
We refer to [Ste51] Lemma 2.8 for a criteria whether the equivalence classes of two (not necessarily equivalent) trivializations for the same bundle projection are isomorphic (see Definition A.4.4) bundles.

**A.1.8 Example** (*"The" Trivial Bundle*)
Let $E$ be a Banach space and $\mathcal{E} := \Omega \times E$. Furthermore, define the natural projection $p : \mathcal{E} \to \Omega$ by $p(x,b) := x$. For each $x \in \Omega$, we endow $p^{-1}(\{x\}) = \{x\} \times E$ with a Banach space structure by identifying it with $E$. It is easy to show, that norm induced topology coincides with the one induced from the product topology. Clearly, $\{\mathrm{Id} : \mathcal{E} \to \Omega\times E\}$ is a trivialization (with associated trivializing cover $\{\Omega\}$). Thus its equivalence class is a bundle. △

**A.1.9 Proposition** (*Refinement of Trivialization*)
Let $\langle \mathcal{E} \overset{p}{\succ} \Omega\rangle$ be a bundle and $\{\phi^{(\lambda)} : \mathcal{E}_{|U_\lambda} \to U_\lambda \times B_\lambda\}_{\lambda\in\Lambda} \in \langle \mathcal{E} \overset{p}{\succ} \Omega\rangle$. Furthermore, let $\{V_\kappa\}_{\kappa\in K}$ be an open cover of $\Omega$ that is finer than $\{U_\lambda\}_{\lambda\in\Lambda}$, i. e. for all $\kappa \in K$ there exists a $\lambda_\kappa \in \Lambda$ such that $V_\kappa \subset U_{\lambda_\kappa}$.
Then $\{\phi^{(\lambda_\kappa)}_{|\mathcal{E}_{|V_\kappa}}\}_{\kappa\in K} \in \langle \mathcal{E} \overset{p}{\succ} \Omega\rangle$. In particular, if $\phi : \mathcal{E}_{|U} \to U \times B$ is a trivializing map, then the restriction $\phi_{|V}$ to any open set $V \overset{\circ}{\subset} U$ is one, too.

*Proof.*
We set $\psi^{(\kappa)} := \phi^{(\lambda_\kappa)}_{|\mathcal{E}_{|V_\kappa}}$ for each $\kappa \in K$. Since $\left(\phi^{(\lambda_\kappa)}\right)^{-1}(V_\kappa \times B_{\lambda_\kappa}) = p^{-1}(V_\kappa) = \mathcal{E}_{|V_\kappa}$ we get indeed that $\psi^{(\kappa)}$ is a homeomorphism from $\mathcal{E}_{|V_\kappa}$ to $V_\kappa \times B_{\lambda_\kappa}$. Clearly, conditions (a) and (b) of Definition A.1.1 hold. Finally, if $(W_i, \chi^{(i)}) \in \{(V_\kappa, \psi^{(\kappa)})\}_{\kappa\in K} \cup \{(U_\lambda, \phi^{(\lambda)})\}_{\lambda\in\Lambda}$ for each $i = 1,2$ with $W := W_1 \cap W_2 \neq \emptyset$ then, by definition, there exist $\lambda_{1,2} \in \Lambda$ such that $\chi^{(i)} = \phi^{(\lambda_i)}_{|\mathcal{E}_{|W_i}}$ and $W \subset$

$U_{\lambda_1} \cap U_{\lambda_2} \neq \emptyset$. Thus the map $W \ni x \mapsto \chi_x^{(2)} \circ (\chi_x^{(1)})^{-1} \in \mathscr{L}(B_{\lambda_1}, B_{\lambda_2})$ is analytic since it coincides with restriction of the analytic map $U_{\lambda_1} \cap U_{\lambda_2} \ni x \mapsto \phi_x^{(\lambda_2)} \circ (\phi_x^{(\lambda_1)})^{-1} \in \mathscr{L}(B_{\lambda_1}, B_{\lambda_2})$ to $W$. Therefore the transition functions of $\{\psi^{(\kappa)}\}_{\kappa \in K}$ alone and joined with $\{\phi^{(\lambda)}\}_{\lambda \in \Lambda}$ fulfill condition A.1.1 (c). □

**A.1.10 Proposition** (*Common Refinement*)
Let $\langle \mathcal{E} \overset{p}{\succ} \Omega \rangle$ and $\langle \mathcal{F} \overset{q}{\succ} \Omega \rangle$ be bundles over the same base space $\Omega$.
Then there are representants $\{\phi^{(\vartheta)}\}_{\vartheta \in \Theta}$ and $\{\psi^{(\vartheta)}\}_{\vartheta \in \Theta}$ of $\langle \mathcal{E} \overset{p}{\succ} \Omega \rangle$ and $\langle \mathcal{F} \overset{q}{\succ} \Omega \rangle$, resp. with the same associated trivializing cover $\{W_\vartheta\}_{\vartheta \in \Theta}$ of $\Omega$ such that for each $\vartheta \in \Theta$ $W_\vartheta$ is connected.

*Proof.*
Let $\{\phi^{(\lambda)}\}_{\lambda \in \Lambda}$ and $\{\psi^{(\kappa)}\}_{\kappa \in K}$ be representants with associated trivializing cover $\{U_\lambda\}_{\lambda \in \Lambda}$ and $\{V_\kappa\}_{\kappa \in K}$ for $\langle \mathcal{E} \overset{p}{\succ} \Omega \rangle$ and $\langle \mathcal{F} \overset{q}{\succ} \Omega \rangle$, resp.. $\{U_\lambda \cap V_\kappa\}_{(\lambda,\kappa) \in \Lambda \times K}$ is an open cover that is finer than both $\{U_\lambda\}_{\lambda \in \Lambda}$ and $\{V_\kappa\}_{\kappa \in K}$. Thus, by Proposition A.1.9, $\{\phi^{(\lambda)}_{|U_\lambda \cap V_\kappa}\}_{(\lambda,\kappa) \in \Lambda \times K}$ and $\{\psi^{(\kappa)}_{|U_\lambda \cap V_\kappa}\}_{(\lambda,\kappa) \in \Lambda \times K}$ are trivializations for $\langle \mathcal{E} \overset{p}{\succ} \Omega \rangle$ and $\langle \mathcal{F} \overset{q}{\succ} \Omega \rangle$, resp.. Furthermore, since $\Omega$ is locally connected[1] by Proposition A.1.9 we can also assume w. l. o. g. that for each $\vartheta \in \Theta$ $W_\vartheta$ is connected. □

**A.1.11 Proposition** (*Exchange of the Banach Space Structure on $B_\lambda$*)
Let $\{\phi^{(\lambda)} : \mathcal{E}_{|U_\lambda} \to U_\lambda \times B_\lambda\}_{\lambda \in \Lambda}$ be a trivialization for a bundle projection $p : \mathcal{E} \to \Omega$. Furthermore, let $\iota_\lambda : B_\lambda \to \tilde{B}_\lambda$ be an isomorphism (of Banach spaces) for every $\lambda \in \Lambda$ and set $i_\lambda := (\mathrm{Id}_{U_\lambda}, \iota_\lambda) : U_\lambda \times B_\lambda \to U_\lambda \times \tilde{B}_\lambda$.
Then $\{i_\lambda \circ \phi^{(\lambda)} : \mathcal{E}_{|U_\lambda} \to U_\lambda \times \tilde{B}_\lambda\}_{\lambda \in \Lambda}$ is a trivialization for $p$, that is equivalent to $\{\phi^{(\lambda)}\}_{\lambda \in \Lambda}$.

*Proof.*
Clearly, $i_\lambda$ and thus $i_\lambda \circ \phi^{(\lambda)}$ are homeomorphic. Condition (a) of Definition A.1.1 obviously holds. Furthermore, for every $\lambda \in \Lambda$ and $x \in U_\lambda$ the induced isomorphism $(i_\lambda \circ \phi^{(\lambda)})_x$ is given by $\mathcal{E}_x \xrightarrow{\phi_x^{(\lambda)}} B_\lambda \xrightarrow{\iota_\lambda} \tilde{B}_\lambda$ and thus is isomorphic. Therefore, condition A.1.1 (b) holds and for all $\lambda, \kappa \in \Lambda$ with $x \in U_\lambda \cap U_\kappa \neq \emptyset$ the transition function is given by $(\iota_\kappa \circ \phi_x^{(\kappa)}) \circ (\iota_\lambda \circ \phi_x^{(\lambda)})^{-1} = \iota_\kappa \circ \left(\phi_x^{(\kappa)} \circ (\phi_x^{(\lambda)})^{-1}\right) \circ (\iota_\lambda)^{-1}$ and thus by Fact 1.5.10 is analytic. We conclude, that condition A.1.1 (c) holds and that $\{i_\lambda \circ \phi^{(\lambda)}\}_{\lambda \in \Lambda}$ is a trivialization

[1]Cf. [Que01] Definition 4.16.

for $p$. It is equivalent to $\{\phi^{(\lambda)}\}_{\lambda\in\Lambda}$ since analogously to the previous argument, all additional transition functions that arise in their union have the form $\left(\phi_x^{(\kappa)}\circ(\phi_x^{(\lambda)})^{-1}\right)\circ(\iota_\lambda)^{-1}$ or $\iota_\kappa\circ\left(\phi_x^{(\kappa)}\circ(\phi_x^{(\lambda)})^{-1}\right)$ and are therefore analytic. □

**A.1.12 Construction** (*Exchange of the Banach Space Structure on $\mathcal{E}_x$*)
Let $\langle\mathcal{E}\overset{p}{\succ}\Omega\rangle$ be a bundle. Furthermore, let $\iota_x:\mathcal{E}_x\to\tilde{E}_x$ be an isomorphism (of Banach spaces) for every $x\in\Omega$.

We set $\tilde{\mathcal{E}}:=\bigcup_{x\in\Omega}\{x\}\times\tilde{E}_x$ and define $i:\mathcal{E}\to\tilde{\mathcal{E}}$ by $i(e):=(p(e),\iota_{p(e)}(e))$ for each $e\in\mathcal{E}$. Clearly, $i$ is bijective. We endow $\tilde{\mathcal{E}}$ with the final topology w. r. t. $i$, i. e. $O\subset\tilde{\mathcal{E}}$ is open iff $i^{-1}(O)$ is open in $\mathcal{E}$ (see [Que01] Satz 3.16). Then $i$ is a homeomorphism.

Denote by $\tilde{p}:\tilde{\mathcal{E}}\to\Omega$ the natural projection $\tilde{p}(x,\tilde{e}):=x$. Clearly, $\tilde{p}\circ i=p$ and thus $\tilde{p}$ is surjective and continuous.

Furthermore, for each $x\in\Omega$ we equip $\tilde{\mathcal{E}}_x=\tilde{p}^{-1}(\{x\})=i(p^{-1}(\{x\}))=i(\mathcal{E}_x)=\{x\}\times\tilde{E}_x$ with a Banach space structure by the natural identification with $\tilde{E}_x$. Then its topology coincides with the induced topology from $\tilde{\mathcal{E}}$: If $\tilde{o}:=\{x\}\times\tilde{o}_x\overset{\circ}{\subset}\{x\}\times\tilde{E}_x$ is open in the Banach space norm topology, then $\iota_x^{-1}(\tilde{o}_x)\overset{\circ}{\subset}\mathcal{E}_x$ is open (in the Banach space norm topology). Thus there exists $O\overset{\circ}{\subset}\mathcal{E}$ such that $\iota_x^{-1}(\tilde{o}_x)=\mathcal{E}_x\cap O$. Setting $\tilde{O}:=i(O)\overset{\circ}{\subset}\tilde{\mathcal{E}}$ we obtain $\tilde{o}=i(\iota_x^{-1}(\tilde{o}_x))=i(\mathcal{E}_x\cap O)=i(\mathcal{E}_x)\cap i(O)=(\{x\}\times\tilde{E}_x)\cap\tilde{O}$ and thus $\tilde{o}$ is open in in the induced topology. Conversely, if we assume that $\emptyset\neq\tilde{O}\overset{\circ}{\subset}\tilde{\mathcal{E}}$, then $i^{-1}(\tilde{O})\overset{\circ}{\subset}\mathcal{E}$ and $o:=i^{-1}(\tilde{O})\cap\mathcal{E}_x\overset{\circ}{\subset}\mathcal{E}_x$. Therefore, $\tilde{O}\cap(\{x\}\times\tilde{E}_x)=i\big(i^{-1}\big(\tilde{O}\cap(\{x\}\times\tilde{E}_x)\big)\big)=\{x\}\times\iota_x\big(i^{-1}\big(\tilde{O}\cap(\{x\}\times\tilde{E}_x)\big)\big)=\{x\}\times\iota_x\big(i^{-1}(\tilde{O})\cap\mathcal{E}_x\big)=\{x\}\times\iota_x(o)$ is open in the Banach space norm topology.

Clearly, for all $x\in\Omega$ the restriction $i_{|\mathcal{E}_x}$ is given by $i_{|\mathcal{E}_x}:\mathcal{E}_x\xrightarrow{\iota_x}\tilde{E}_x\xrightarrow{\cong}\{x\}\times E_x\xrightarrow{=}\tilde{\mathcal{E}}_x$ and therefore is an isomorphism from $\mathcal{E}_x$ to $\tilde{\mathcal{E}}_x$.

Next, for a representant $\{\phi^{(\lambda)}:\mathcal{E}_{|U_\lambda}\to U_\lambda\times B_\lambda\}_{\lambda\in\Lambda}$ of $\langle\mathcal{E}\overset{p}{\succ}\Omega\rangle$ define $\tilde{\phi}^{(\lambda)}:=\phi^{(\lambda)}\circ i^{-1}$ for each $\lambda\in\Lambda$. Then $\{\tilde{\phi}^{(\lambda)}\}_{\lambda\in\Lambda}$ is a trivialization for $\tilde{p}$: Since $i^{-1}:\mathcal{E}_{|U_\lambda}\to\tilde{\mathcal{E}}_{|U_\lambda}$ is homeomorphic, $\tilde{\phi}^{(\lambda)}:\tilde{\mathcal{E}}_{|U_\lambda}\to U_\lambda\times B_\lambda$ is a homeomorphism. By construction, condition (a) of Definition A.1.1 holds. Furthermore for each $\lambda\in\Lambda$ and $x\in U_\lambda$, $\tilde{\phi}_x^{(\lambda)}$ is given by $\tilde{\mathcal{E}}_x\xrightarrow{(i_{|\mathcal{E}_x})^{-1}}\mathcal{E}_x\xrightarrow{\phi_x^{(\lambda)}}B_\lambda$ and thus is an isomorphism. Therefore, condition (b) holds. Finally, since for all $\lambda,\kappa\in\Lambda$

with $x \in U_\lambda \cap U_\kappa \neq \emptyset$ the transition function is given by $\tilde{\phi}_x^{(\kappa)} \circ \left(\tilde{\phi}_x^{(\lambda)}\right)^{-1} = \phi_x^{(\kappa)} \circ i^{-1} \circ \left(\phi_x^{(\lambda)} \circ i^{-1}\right)^{-1} = \phi_x^{(\kappa)} \circ \left(\phi_x^{(\lambda)}\right)^{-1}$ and thus is analytic, condition (c) holds.

Therefore the equivalence class $\langle \tilde{\mathcal{E}} \overset{\tilde{p}}{\succ} \Omega \rangle$ of $\{\tilde{\phi}^{(\lambda)}\}_{\lambda \in \Lambda}$ is a bundle.

For each $\xi \in \Omega$ we choose $\lambda \in \Lambda$ such that $\xi \in U_\lambda$. Then the map $\mathcal{I}_x$ : $B_\lambda \overset{\left(\phi_x^{(\lambda)}\right)^{-1}}{\longrightarrow} \mathcal{E}_x \overset{i_{|\mathcal{E}_x}}{\longrightarrow} \tilde{\mathcal{E}}_x \overset{\tilde{\phi}_x^{(\lambda)}}{\longrightarrow} B_\lambda$ coincides with the "constant" $\mathrm{Id}_{B_\lambda}$ and thus is analytically depending on $x \in U_\lambda$. We will later say (see Definition A.4.4 and Proposition A.4.5 (b)): "The bundles $\langle \mathcal{E} \overset{p}{\succ} \Omega \rangle$ and $\langle \tilde{\mathcal{E}} \overset{\tilde{p}}{\succ} \Omega \rangle$ are isomorphic". △

**A.1.13 Proposition** (*Isomorphism of Fibers*)
Let $\langle \mathcal{E} \overset{p}{\succ} \Omega \rangle$ be a bundle.
Then on every connected component $C$ of $\Omega$, all fibers $\mathcal{E}_x$ for all $x \in C$ are isomorphic (as Banach spaces).

If, in particular, the base space $\Omega$ is connected, then all fibers $\mathcal{E}_x$ for all $x \in \Omega$ are isomorphic (as Banach spaces).

*Proof.*
Let $C$ be a connected component of $\Omega$ and $\{\phi^{(\lambda)} : \mathcal{E}_{|U_\lambda} \to U_\lambda \times B_\lambda\}_{\lambda \in \Lambda}$ a representant of $\langle \mathcal{E} \overset{p}{\succ} \Omega \rangle$. Furthermore, let $x, y \in C$. Since $C$ is connected there exists $\lambda_1, \ldots, \lambda_n \in \Lambda$ such that $x \in U_{\lambda_1}$, $y \in U_{\lambda_n}$ and $U_{\lambda_i} \cap U_{\lambda_j} \neq \emptyset$ for all $|i - j| \leq 1$ (see e.g. [Que01] Lemma 4.8). Thus $B_{\lambda_i} \cong \mathcal{E}_\xi \cong B_{\lambda_j}$ for some $\xi \in U_{\lambda_i} \cap U_{\lambda_j}$ and all $|i-j| \leq 1$. Therefore $\mathcal{E}_x \cong B_{\lambda_1} \cong B_{\lambda_2} \cong \cdots \cong B_{\lambda_n} \cong \mathcal{E}_y$. □

**A.1.14 Definition**
Let $\langle \mathcal{E} \overset{p}{\succ} \Omega \rangle$ be a bundle and $E$ a Banach space.

We say $\langle \mathcal{E} \overset{p}{\succ} \Omega \rangle$ is a *bundle with fiber* $E$ if all fibers $\mathcal{E}_x$ for all $x \in \Omega$ are isomorphic (as Banach spaces) to $E$.

## A.2 Sections

Throughout this section let $\langle \mathcal{E} \overset{p}{\succ} \Omega \rangle$ be a bundle. We denote the Banach space structure on the fibers by $(\mathcal{E}_x, +_x : \mathcal{E}_x \times \mathcal{E}_x \to \mathcal{E}_x, \cdot_x : \mathbb{C} \times \mathcal{E}_x \to \mathcal{E}_x, \|\cdot\|_x)$ for each $x \in \Omega$. Furthermore, let $\emptyset \neq O \overset{\circ}{\subset} \Omega$.

**A.2.1 Definition** (*Section*)
A function $s : O \to \mathcal{E}$ is called *(analytic) section (of $\langle \mathcal{E} \overset{p}{\succ} \Omega \rangle$) (over $O$)* if $p \circ s = \mathrm{Id}_O$ and for every $\xi \in O$ there is a trivializing map $\phi : \mathcal{E}_{|U} \to U \times B$ for $\langle \mathcal{E} \overset{p}{\succ} \Omega \rangle$ such that $\xi \in U$ and[2] $\big[x \mapsto \phi_x\big(s(x)\big)\big] \in A(U \cap O, B)$. We denote the set of all sections of $\langle \mathcal{E} \overset{p}{\succ} \Omega \rangle$ over $O$ by $\Gamma(O, \langle \mathcal{E} \overset{p}{\succ} \Omega \rangle)$.

Throughout the rest of this section we set $\Gamma := \Gamma(O, \langle \mathcal{E} \overset{p}{\succ} \Omega \rangle)$.

**A.2.2 Remark**
Let $\emptyset \neq o \overset{\circ}{\subset} O$.
Then $s \in \Gamma(O, \langle \mathcal{E} \overset{p}{\succ} \Omega \rangle))$ obviously implies $s_{|o} \in \Gamma(o, \langle \mathcal{E} \overset{p}{\succ} \Omega \rangle))$.

**A.2.3 Proposition**
Let $[s : O \to \mathcal{E}] \in \Gamma$.
Then $s$ is continuous.

*Proof.*
Let $\xi \in O$. Furthermore, let $\phi : \mathcal{E}_{|U} \to U \times B$ be a trivializing map for $\langle \mathcal{E} \overset{p}{\succ} \Omega \rangle$ such that $\xi \in U$ and $\big[x \mapsto \phi_x\big(s(x)\big)\big] \in A(U \cap O, B)$. Thus by Fact 1.5.1 $\big[x \mapsto \big(x, \phi_x\big(s(x)\big)\big)\big] \in C(U \cap O, (U \cap O) \times B)$. Therefore $\big[x \mapsto \phi^{-1}\big(x, \phi_x\big(s(x)\big)\big)\big] \in C(U \cap O, \mathcal{E}_U)$. On the other hand $\big(x, \phi_x\big(s(x)\big)\big) = \big(p(s(x)), \phi_x\big(s(x)\big)\big) = \phi\big(s(x)\big)$ and thus $\phi^{-1}\big(\phi\big(s(x)\big)\big) = s(x)$ for all $x \in U \cap O$. This yields that $s$ is continuous in $\xi$ and therefore continuous on $O$.

**A.2.4 Proposition**
In the situation of Definition A.2.1, $s \in \Gamma$ iff $p \circ s = \mathrm{Id}_O$ and $\big[x \mapsto \psi_x\big(s(x)\big)\big] \in A(V \cap O, C)$ for every trivializing map $\psi : \mathcal{E}_{|V} \to V \times C$ for $\langle \mathcal{E} \overset{p}{\succ} \Omega \rangle$.

*Proof.*
The direction "$\Longleftarrow$" is clear. In order to show "$\Longrightarrow$", let $\xi \in V \cap O$. Then there is a trivializing map $\phi : \mathcal{E}_{|U} \to U \times B$ for $\langle \mathcal{E} \overset{p}{\succ} \Omega \rangle$ such that $\xi \in U$ and $\big[x \mapsto \phi_x\big(s(x)\big)\big] \in A(U \cap O, B)$. Then for each $x \in U \cap V \cap O$ $\psi_x\big(s(x)\big) = \big(\psi_x \circ (\phi_x)^{-1} \circ \phi_x\big)\big(s(x)\big) = \big(\psi_x \circ (\phi_x)^{-1}\big)\big(\phi_x\big(s(x)\big)\big)$. Since $\big[x \mapsto \psi_x \circ (\phi_x)^{-1}\big] \in A(U \cap V \cap O, \mathscr{L}(B, C)$ by definition, this yields $\big[x \mapsto \psi_x\big(s(x)\big)\big] \in A(U \cap V \cap O, C)$ (cf. [Cha85] Example 3.10, Exercise 5D and Theorem 14.13). In

[2] We note that $x \mapsto \phi_x\big(s(x)\big)$ is well-defined since $s(x) \in \mathcal{E}_x$ for all $x \in O$.

particular $x \mapsto \psi_x\big(s(x)\big)$ is analytic in the point $\xi$ and thus analytic on $V \cap O$. □

**A.2.5 Definition and Proposition** (*Vector Space Structure*)
For all $s, \tilde{s} \in \Gamma$, $\alpha \in \mathbb{C}$ and $x \in \Omega$ we define $(\alpha \cdot s)(x) := (\alpha s)(x) := \alpha \cdot_x s(x)$ and $(s + \tilde{s})(x) := s(x) +_x \tilde{s}(x)$ for all $x \in \Omega$.
Then $\alpha \cdot s, s + \tilde{s} \in \Gamma$ and $(\Gamma, +, \cdot)$ is a vector space.

*Proof.*
It is obvious that $(\Gamma, +, \cdot)$ is a vector space once we have shown that $\alpha \cdot s, s + \tilde{s} \in \Gamma$ for all $s, \tilde{s} \in \Gamma$ and $\alpha \in \mathbb{C}$. Let $\psi : \mathcal{E}_{|U} \to U \times B$ be a trivializing map for $\langle \mathcal{E} \overset{p}{\succ} \Omega \rangle$. Then by Proposition A.2.4 $\big[x \mapsto \psi_x\big(s(x)\big)\big], \big[x \mapsto \psi_x\big(\tilde{s}(x)\big)\big] \in A(U \cap O, B)$. In addition, since $\psi_x$ is an Banach space isomorphism, $\psi_x\big((\alpha \cdot s)(x)\big) = \psi_x\big(\alpha \cdot_x s(x)\big) = \alpha \psi_x\big(s(x)\big)$ and $\psi_x\big((s + \tilde{s})(x)\big) = \psi_x\big(s(x) +_x \tilde{s}(x)\big) = \psi_x\big(s(x)\big) + \psi_x\big(\tilde{s}(x)\big)$ for all $x \in U \cap O$. Thus by Fact 1.5.6 $\big[x \mapsto \psi_x\big((\alpha \cdot s)(x)\big)\big], \big[x \mapsto \psi_x\big((s + \tilde{s})(x)\big)\big] \in A(U \cap O, B)$ and then by Proposition A.2.4 $\alpha \cdot s, s + \tilde{s} \in \Gamma$. □

**A.2.6 Definition**
We endow $\Gamma$ with the compact-open topology, i. e. the topology generated by the subbasis of all sets of the form $B_\Gamma(K, \mathcal{O}) := \{\, s \in \Gamma : s(K) \subset \mathcal{O} \,\}$ where $K \subset\subset O$ and $\mathcal{O} \overset{\circ}{\subset} \mathcal{E}$.

**A.2.7 Proposition** (*Fréchet Space Structure*)
Let $\{\phi^{(\lambda)} : \mathcal{E}_{|U_\lambda} \to U_\lambda \times B_\lambda\}_{\lambda \in \Lambda} \in \langle \mathcal{E} \overset{p}{\succ} \Omega \rangle$ and $\mathcal{K} := \{\, (\lambda, K) : \lambda \in \Lambda, \emptyset \neq K \subset\subset U_\lambda \cap O \,\}$.
For each $(\lambda, K) \in \mathcal{K}$ we define $p_{(\lambda,K)} : \Gamma \to \mathbb{R}$ by

$$p_{(\lambda,K)}(s) := \sup_{x \in K} \big\| \phi_x^{(\lambda)}\big(s(x)\big) \big\|_{B_\lambda}.$$

Then $p_{(\lambda,K)}$ is a seminorm. The topology induced by the family of seminorms[3] $\{p_\kappa\}_{\kappa \in \mathcal{K}}$ coincides with the compact-open topology on $\Gamma$.
$\Gamma$ is a (complex) Fréchet space.

*Proof.*
Since $O$ is (a metric space and thus) paracompact and separable, $O$ is a Lindelöf space (see [Dug70] Definition VIII.6.4 and the theorem in [Dug70] VIII.7.4). Therefore there exists a countable subset $\Lambda_{\text{cnt}} \subset \Lambda$ such that $\{U_\lambda\}_{\lambda \in \Lambda_{\text{cnt}}}$ covers

[3] Cf. [Jän05] Section 2.5.

$O$. For each $\lambda \in \Lambda_{\text{cnt}}$, analogously $U_\lambda$ is a locally compact Lindelöf space and therefore countable at infinity, i. e. there is a countable family $\{K_n^{(\lambda)}\}_{n\in\mathbb{N}}$ such that for every $K \subset\subset U_\lambda$ there is $n \in \mathbb{N}$ with $K \subset K_n^{(\lambda)}$. Then $\mathcal{K}_{\text{cnt}} := \bigcup_{\lambda\in\Lambda_{\text{cnt}}} \left(\{\lambda\} \times \{K_n^{(\lambda)}\}_{n\in\mathbb{N}}\right)$ is a countable subset of $\mathcal{K}$.

It is clear that $p_\kappa$ is a seminorm for each $\kappa \in \mathcal{K}$.

We denote the topology induced by the family of seminorms[4] $\{p_\kappa\}_{\kappa\in\mathcal{K}}$ and $\{p_\kappa\}_{\kappa\in\mathcal{K}_{\text{cnt}}}$ by $\tau$ and $\tau_{\text{cnt}}$, resp.. Furthermore, for all $\kappa \in \mathcal{K}$, $\epsilon > 0$ and $s \in \Gamma$ we denote by $B_\epsilon^\kappa(s) := \{\sigma \in \Gamma : p_\kappa(s - \sigma) < \epsilon\}$ the open balls of $\tau$ (and particularly of $\tau_{\text{cnt}}$ if $\kappa \in \mathcal{K}_{\text{cnt}}$).

Let $K \subset\subset O$ and $\mathcal{O} \overset{\circ}{\subset} \mathcal{E}$. We will now show that $B_\Gamma(K, \mathcal{O}) \in \tau_{\text{cnt}}$. Therefore, w. l. o. g. we can assume that there exists $s \in B_\Gamma(K, \mathcal{O})$ (and thus $K \subset p(\mathcal{O})$) and that there exists $x \in K$. Then there is $\lambda_x \in \Lambda_{\text{cnt}}$ with $x \in U_{\lambda_x}$. Then $\left(x, \phi_x^{(\lambda_x)}(s(x))\right) = \phi^{(\lambda_x)}(s(x)) \in \phi^{(\lambda_x)}(\mathcal{E}_{|U_{\lambda_x}} \cap \mathcal{O}) \overset{\circ}{\subset} U_{\lambda_x} \times B_{\lambda_x}$. Thus there are $o_x \overset{\circ}{\subset} U_{\lambda_x}$ and $\epsilon_x > 0$ such that $\left(x, \phi_x^{(\lambda_x)}(s(x))\right) \in o_x \times b_{\epsilon_x}^{(x)} \overset{\circ}{\subset} \phi^{(\lambda_x)}(\mathcal{E}_{|U_{\lambda_x}} \cap \mathcal{O})$, where $b_\epsilon^{(x)} := B_{B_{\lambda_x}}\left(\phi_x^{(\lambda_x)}(s(x)), \epsilon\right)$ for each $\epsilon > 0$. By continuity of $\phi^{(x_0)} \circ s_{|U_{\lambda_x}\cap O}$ we can additionally choose $o_x$ "small enough" such that $o_x \subset O$ and $\phi_\xi^{(\lambda_x)}(s(\xi)) \in b_{\epsilon_x/2}^{(x)}$ for all $\xi \in o_x$.

$\{o_x\}_{x\in K}$ is an open cover of $K$. By compactness of $K$ and, since $K$ is a metric and thus normal space, by [Que01] Satz 7.12 and [Que01] Satz 8.6 there is a finite set $K_{\text{fin}} \subset K$ and $k_x \subset\subset o_x$ for each $x \in K_{\text{fin}}$ such that $\{k_x\}_{x\in K_{\text{fin}}}$ covers $K$. For each $x \in K_{\text{fin}}$ there is $K_x \subset\subset U_{\lambda_x}$ such that $(\lambda_x, K_x) \in \mathcal{K}_{\text{cnt}}$ and $k_x \subset K_x$. We set $\epsilon := 1/2 \min\{\epsilon_x : x \in K_{\text{fin}}\} > 0$ and $B_s := \bigcap_{x\in K_{\text{fin}}} B_\epsilon^{(\lambda_x, K_x)}(s)$.

We now show that $B_s \subset B_\Gamma(K, \mathcal{O})$: Let $\sigma \in B_s$ and $\xi \in K$. Then there is $x \in K_{\text{fin}}$ such that $\xi \in k_x$. Since $\sigma \in B_\epsilon^{(\lambda_x, K_x)}(s)$ we obtain $\left\|\phi_\xi^{(\lambda_x)}(\sigma(\xi)) - \phi_\xi^{(\lambda_x)}(s(\xi))\right\|_{B_{\lambda_x}} = \left\|\phi_\xi^{(\lambda_x)}((\sigma - s)(\xi))\right\|_{B_{\lambda_x}} < \epsilon \leq \epsilon_x/2$. Also, $\xi \in o_x$ and thus $\phi_\xi^{(\lambda_x)}(s(\xi)) \in b_{\epsilon_x/2}^{(x)}$. Therefore $\|\phi_\xi^{(\lambda_x)}(s(\xi)) - \phi_x^{(\lambda_x)}(s(x))\|_{B_{\lambda_x}} < \epsilon_x/2$. Thus $\|\phi_\xi^{(\lambda_x)}(\sigma(\xi)) - \phi_x^{(\lambda_x)}(s(x))\|_{B_{\lambda_x}} < \epsilon_x$, or reformulated, $\phi_\xi^{(\lambda_x)}(\sigma(\xi)) \in b_{\epsilon_x}^{(x)}$. Therefore $\phi^{(\lambda_x)}(\sigma(\xi)) = \left(\xi, \phi_\xi^{(\lambda_x)}(\sigma(\xi))\right) \in \phi^{(\lambda_x)}(\mathcal{E}_{|U_{\lambda_x}} \cap \mathcal{O})$. This yields $\sigma(\xi) \in \mathcal{O}$ and thus $\sigma \in B_\Gamma(K, \mathcal{O})$.

Since trivially $s \in B_s$ and $B_s$ is open in $\tau_{\text{cnt}}$, we conclude that $B_\Gamma(K, \mathcal{O})$ consists only of interior points w. r. t. $\tau_{\text{cnt}}$ and therefore is open in $\tau_{\text{cnt}}$. Thus

[4] Cf. [Jän05] Section 2.5.

the compact-open topology is coarser than $\tau_{\text{cnt}}$.

Of course, $\tau_{\text{cnt}}$ is coarser than $\tau$.

Finally, let $(\lambda, K) \in \mathcal{K}$, $\epsilon > 0$ and $s \in \Gamma$. We set $o := \bigcup_{\xi \in U_\lambda \cap O} \big(\{\xi\} \times B_{B_\lambda}(\phi_\xi^{(\lambda)}(s(\xi)), \epsilon)\big)$. We note that $U_\lambda \times B_\lambda \ni (\xi, \beta) \mapsto \big(\xi, \beta + \phi_\xi^{(\lambda)}(s(\xi))\big) \in U_\lambda \times B_\lambda$ is a homeomorphism and $o$ is the image of $(U_\lambda \cap O) \times B_{B_\lambda}(0, \epsilon) \overset{\circ}{\subset} U_\lambda \times B_\lambda$ under that homeomorphism. Therefore $o \overset{\circ}{\subset} U_\lambda \times B_\lambda$. Then $(\phi^{(\lambda)})^{-1}(o) \overset{\circ}{\subset} \mathcal{E}$ and $B_\Gamma(K, (\phi^{(\lambda)})^{-1}(o)) = \{\sigma \in \Gamma : \sigma(\xi) \in (\phi^{(\lambda)})^{-1}(o) \text{ for all } \xi \in K\} = \{\sigma \in \Gamma : \phi^{(\lambda)}(\sigma(\xi)) \in o \text{ for all } \xi \in K\} = \{\sigma \in \Gamma : \phi_\xi^{(\lambda)}(\sigma(\xi)) \in B_{B_\lambda}(\phi_\xi^{(\lambda)}(s(\xi)), \epsilon) \text{ for all } \xi \in K\} = \{\sigma \in \Gamma : \|\phi_\xi^{(\lambda)}((s - \sigma)(\xi))\|_{B_\lambda} < \epsilon \text{ for all } \xi \in K\}$. We note that $\{\sigma \in \Gamma : \|\phi_\xi^{(\lambda)}((s - \sigma)(\xi))\|_{B_\lambda} < \epsilon \text{ for all } \xi \in K\} = B_\epsilon^{(\lambda,K)}(s)$ since the continuous function $K \ni \xi \mapsto \|\phi_\xi^{(\lambda)}((s - \sigma)(\xi))\|_{B_\lambda}$ attains its supremum on $K$. Therefore $B_\epsilon^{(\lambda,K)}(s) = B_\Gamma(K, (\phi^{(\lambda)})^{-1}(o))$ is open in the compact-open topology. Thus $\tau$ is coarser than the compact-open topology.

We conclude that $\tau$, $\tau_{\text{cnt}}$ and the compact-open topology all coincide.

In view of $\tau$ it is clear that $\Gamma$ is a Hausdorff space and in view of $\tau_{\text{cnt}}$ it then follows that $\Gamma$ is a pre-Fréchet space.

Before we show completeness we treat the special case where $O$ belongs to the trivializing cover:

### A.2.8 Proposition

Let $\phi : \mathcal{E}_{|U} \to U \times B$ be a trivializing map for $\langle \mathcal{E} \overset{p}{\succ} \Omega \rangle$.

Then[5] $s \mapsto [x \mapsto \phi_x(s(x))] \in \mathscr{L}(\Gamma(U, \langle \mathcal{E} \overset{p}{\succ} \Omega \rangle), A(U, B))$ is an isomorphism.

*Proof.*

For all $s \in \Gamma(U, \langle \mathcal{E} \overset{p}{\succ} \Omega \rangle)$ we denote by $T(s) := [x \mapsto \phi_x(s(x))]$; then by Proposition A.2.4 $T(s) \in A(U, B)$ and clearly $T : \Gamma(U, \langle \mathcal{E} \overset{p}{\succ} \Omega \rangle) \to A(U, B)$ is linear and injective. If conversely $f \in A(U, B)$, then obviously $s_f := [U \ni x \mapsto (\phi_x)^{-1}(s(x))] \in \Gamma(U, \langle \mathcal{E} \overset{p}{\succ} \Omega \rangle)$ and $T(s_f) = f$. Thus $T$ is surjective. The open balls in $\Gamma(U, \langle \mathcal{E} \overset{p}{\succ} \Omega \rangle)$ and $A(U, B)$ are given by $B_\epsilon^K(s) := \{\sigma \in \Gamma :$

---

[5] We have introduced the symbol $\mathscr{L}(\dots)$ only for complete spaces, but it is clear from the statement itself that $\Gamma(U, \langle \mathcal{E} \overset{p}{\succ} \Omega \rangle)$ is indeed complete.

$\sup\limits_{x\in K} \left\|\phi_x\big((s-\sigma)(x)\big)\right\|_B < \epsilon\}$ and $\tilde{B}^K_\epsilon(f) := \{\,\tilde{f} \in A(U,B) : \sup_{x\in K} \|(f-\tilde{f})(x)\|_B < \epsilon\}$, resp., where $s \in \Gamma(U, \langle \mathcal{E} \overset{p}{\succ} \Omega\rangle)$, $f \in A(U,B)$, $\emptyset \neq K \subset\subset U$ and $\epsilon > 0$. Then $T(B^K_\epsilon(s)) = \tilde{B}^K_\epsilon(Ts)$ for all $s \in \Gamma(U, \langle \mathcal{E} \overset{p}{\succ} \Omega\rangle)$, $\emptyset \neq K \subset\subset U$ and $\epsilon > 0$. Thus $T$ is continuous and open. □

*Proof of Proposition A.2.7 (Continuation).*
It remains to show completeness.

Let $(s_n)_{n\in\mathbb{N}} \subset \Gamma$ be a Cauchy sequence[6]. Then for each $x \in O$ $(s_n(x))_{n\in\mathbb{N}}$ is a Cauchy sequence in $\mathcal{E}_x$ and thus converge to some $s(x) \in \mathcal{E}_x$. We will now show that the so-defined function $s : O \to \mathcal{E}$ is a section.

$p \circ s = \mathrm{Id}_O$ holds by construction.

Let $\xi \in O$. Then there is $\lambda \in \Lambda$ such that $\xi \in U_\lambda$ (and thus $U_\lambda \cap O \neq \emptyset$). For all $n \in \mathbb{N}$ we set $f^{(\lambda)}_n := T_\lambda((s_n)_{|U_\lambda\cap O}) \in A(U_\lambda \cap O, B_\lambda)$, where $T_\lambda$ is the isomorphism given by Proposition A.2.8 w. r. t. the trivializing map $\phi^{(\lambda)} : \mathcal{E}_{|U_\lambda\cap O} \to (U_\lambda \cap O) \times B_\lambda$ for $\langle \mathcal{E} \overset{p}{\succ} \Omega\rangle$. Proposition A.2.8 yields that $(f^{(\lambda)}_n)_{n\in\mathbb{N}}$ is a Cauchy sequence in $A(U_\lambda \cap O, B_\lambda)$. Thus there exists $f^{(\lambda)} \in A(U_\lambda \cap O, B_\lambda)$ such that $f^{(\lambda)}_n \overset{n\to\infty}{\longrightarrow} f^{(\lambda)}$ In particular $f^{(\lambda)}_n(x) \overset{n\to\infty}{\longrightarrow} f^{(\lambda)}(x)$ in $B_\lambda$ for all $x \in U_\lambda \cap O$. On the other, since $\phi^{(\lambda)}_x \in \mathscr{L}(\mathcal{E}_x, B_\lambda)$ for each $x \in U_\lambda \cap O$, $f^{(\lambda)}_n(x) = \phi^{(\lambda)}_x\big(s_n(x)\big) \overset{n\to\infty}{\longrightarrow} \phi^{(\lambda)}_x\big(s(x)\big)$ in $B_\lambda$ for all $x \in U_\lambda \cap O$. Thus $U_\lambda \cap O \ni x \mapsto \phi^{(\lambda)}_x\big(s(x)\big)$ coincides with $f^{(\lambda)} \in A(U_\lambda \cap O, B_\lambda)$.

Therefore $s$ is a section.

Let $N$ be a neighborhood of $s$ in $\Gamma$. Then there are $\epsilon > 0$, $l \in \mathbb{N}$ and $(\lambda_i, K_i) \in \mathcal{K}$ for each $i = 1, \ldots, l$ such that $\bigcap_{i=1}^{l} B^{(\lambda_i,K_i)}_\epsilon(s) \subset N$. Then for each $i = 1, \ldots, l$ there is $m_i \in \mathbb{N}$ such that $f^{(\lambda_i)}_n \in \tilde{B}^{K_{\lambda_i}}_\epsilon(f^{(\lambda_i)})$ for all $n \geq m_i$, where $B^{K_{\lambda_i}}_\epsilon$ is defined as in the proof of Proposition A.2.8. Since $T_{\lambda_i}(s_{|U_{\lambda_i}\cap O}) = f^{(\lambda_i)}$, from the proof of Proposition A.2.8 we then obtain $s_n \in B^{(\lambda_i,K_i)}_\epsilon(s)$ for all $n \geq m_i$ for each $i = 1, \ldots, l$. Thus $s_n \in N$ for all $n \geq \max\{m_1, \ldots, m_l\}$. Hence $s_n \underset{\Gamma}{\overset{n\to\infty}{\longrightarrow}} s$. □

## A.2.9 Remark

Let $\emptyset \neq o \overset{\circ}{\subset} O$.

Then obviously $[s \mapsto s_{|o}] \in \mathscr{L}(\Gamma, \Gamma(o, \langle \mathcal{E} \overset{p}{\succ} \Omega\rangle))$.

[6] i. e. for every neighborhood $N$ of $0 \in \Gamma$ there is $n_0 \in \mathbb{N}$ such that $s_n - s_m \in N$ for all $n, m \geq n_0$

**A.2.10 Proposition** (*Identity Theorem for Analytic Sections*)
Assume that $O$ is connected and let $s, \tilde{s} \in \Gamma$. If there is $\emptyset \neq o \mathring{\subset} O$ such that $s_{|o} = \tilde{s}_{|o}$ then $s = \tilde{s}$ (on $O$).

*Proof.*
By Proposition A.1.10 we can assume w.l.o.g., that $U_\lambda$ is connected for each $\lambda \in \Lambda$. Let $x \in O$. It suffices to show that $s(x) = \tilde{s}(x)$. Let $y \in o$. Since $O$ is connected there exists $\lambda_1, \ldots, \lambda_n \in \Lambda$ such that $y \in U_{\lambda_1}$, $x \in U_{\lambda_n}$ and $O \cap U_{\lambda_i} \cap U_{\lambda_j} \neq \emptyset$ for all $|i - j| \leq 1$ (see e.g. [Que01] Lemma 4.8). By Proposition A.2.4 $\big[x \mapsto \phi_x^{(\lambda_1)}(s(x))\big], \big[x \mapsto \phi_x^{(\lambda_1)}(\tilde{s}(x))\big] \in A(O \cap U_{\lambda_1}, B_{\lambda_1})$ and $\big[x \mapsto \phi_x^{(\lambda_1)}(s(x))\big]_{|o \cap U_{\lambda_1}} = \big[x \mapsto \phi_x^{(\lambda_1)}(\tilde{s}(x))\big]_{|o \cap U_{\lambda_1}}$. Therefore by the identity theorem for analytic functions (see [Cha85] Theorem 12.9) $\big[x \mapsto \phi_x^{(\lambda_1)}(s(x))\big]_{|O \cap U_{\lambda_1}} = \big[x \mapsto \phi_x^{(\lambda_1)}(\tilde{s}(x))\big]_{|O \cap U_{\lambda_1}}$. Since $\phi_x^{(\lambda_1)}$ is bijective for each $x \in U_{\lambda_1}$ we obtain $s_{|O \cap U_{\lambda_1}} = \tilde{s}_{|O \cap U_{\lambda_1}}$. Similarly $\big[x \mapsto \phi_x^{(\lambda_2)}(s(x))\big], \big[x \mapsto \phi_x^{(\lambda_2)}(\tilde{s}(x))\big] \in A(O \cap U_{\lambda_2}, B_{\lambda_2})$ and $\big[x \mapsto \phi_x^{(\lambda_2)}(s(x))\big]_{|O \cap U_{\lambda_1} \cap U_{\lambda_2}} = \big[x \mapsto \phi_x^{(\lambda_2)}(\tilde{s}(x))\big]_{|O \cap U_{\lambda_1} \cap U_{\lambda_2}}$ and thus $s_{|O \cap U_{\lambda_2}} = \tilde{s}_{|O \cap U_{\lambda_2}}$. Analogously, we obtain $s_{|O \cap U_{\lambda_i}} = \tilde{s}_{|O \cap U_{\lambda_i}}$ for each $i = 3, \ldots, n$. Thus $s(x) = \tilde{s}(x)$. □

## A.3 Restrictions

**A.3.1 Construction and Definition** (*Restriction*)
Let $\langle \mathcal{E} \overset{p}{\succ} \Omega \rangle$ be a bundle and $\emptyset \neq O \mathring{\subset} \Omega$. Furthermore, let $\{\phi^{(\lambda)}\}_{\lambda \in \Lambda}$ be a representant of $\langle \mathcal{E} \overset{p}{\succ} \Omega \rangle$.

We set $\Lambda_{|O} := \{\, \lambda \in \Lambda : O \cap U_\lambda \neq \emptyset \,\}$ and for each $\lambda \in \Lambda_{|O}$ we denote by $\phi_{|O}^{(\lambda)}$ the restriction of $\phi^{(\lambda)}$ to $\mathcal{E}_{|O \cap U_\lambda}$.

Then clearly, $\{O \cap U_\lambda\}_{\lambda \in \Lambda_{|O}}$ is an open cover of $O$, $\phi_{|O}^{(\lambda)} : \mathcal{E}_{|O \cap U_\lambda} \to (O \cap U_\lambda) \times B_\lambda$ is a homeomorphism for each $\lambda \in \Lambda_{|O}$ and $\{\phi_{|O}^{(\lambda)}\}_{\lambda \in \Lambda_{|O}}$ is a trivialization for $p_{|O}$.

Thus equivalence class of $\{\phi_{|O}^{(\lambda)}\}_{\lambda \in \Lambda_{|O}}$ is a bundle and is denoted by $\langle \mathcal{E} \overset{p}{\succ} \Omega \rangle_{|O}$. It's called the *restriction* (of the bundle $\langle \mathcal{E} \overset{p}{\succ} \Omega \rangle$).

**A.3.2 Remark**
Let $\langle \mathcal{E} \overset{p}{\succ} \Omega \rangle$ be a bundle and $\emptyset \neq o \overset{\circ}{\subset} O \overset{\circ}{\subset} \Omega$.
Then obviously $\Gamma(o, \langle \mathcal{E} \overset{p}{\succ} \Omega \rangle))$ coincides with $\Gamma(o, \langle \mathcal{E} \overset{p}{\succ} \Omega \rangle_{|O}))$.

## A.4 Homomorphisms

**A.4.1 Definition** (*Homomorphism*)
Let $\langle \mathcal{E} \overset{p}{\succ} \Omega \rangle$ and $\langle \mathcal{F} \overset{q}{\succ} \Omega \rangle$ be bundles (over the same base space $\Omega$).
A function $A : \mathcal{E} \to \mathcal{F}$ is called *homomorphism* (of bundles), if

(a) the diagram

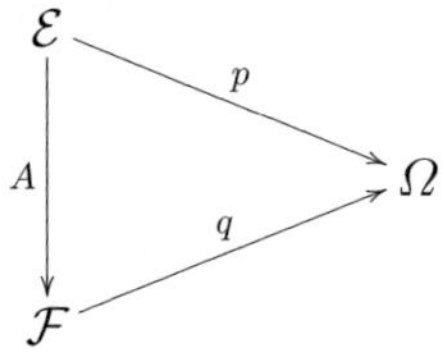


commutes, i. e. $q \circ A = p$ and

(b) for all $x \in \Omega$ the induced operator on the fibers $A_x := A_{|\mathcal{E}_x}$ is linear and bounded, i. e. $A_x \in \mathscr{L}(\mathcal{E}_x, \mathcal{F}_x)$ and

(c) for all $\xi \in \Omega$ there are trivializing maps $\phi : \mathcal{E}_{|U} \to U \times B$ for $\langle \mathcal{E} \overset{p}{\succ} \Omega \rangle$ and $\psi : \mathcal{F}_{|V} \to V \times C$ for $\langle \mathcal{F} \overset{q}{\succ} \Omega \rangle$, resp., with $\xi \in U \subset V$ such that the *trivialized induced map*

$$\mathcal{A}_x : B \overset{(\phi_x)^{-1}}{\longrightarrow} \mathcal{E}_x \overset{A_x}{\longrightarrow} \mathcal{F}_x \overset{\psi_x}{\longrightarrow} C$$

is analytically depending on $x \in U$, i. e. $[x \mapsto \mathcal{A}_x] \in A(U, \mathscr{L}(B, C))$

holds. By abuse of notation we write $A : \langle \mathcal{E} \overset{p}{\succ} \Omega \rangle \to \langle \mathcal{F} \overset{q}{\succ} \Omega \rangle$. We say the homomorphism $A : \langle \mathcal{E} \overset{p}{\succ} \Omega \rangle \to \langle \mathcal{F} \overset{q}{\succ} \Omega \rangle$ is *injective* or *bijective* if the map $A : \mathcal{E} \to \mathcal{F}$ is injective or bijective, resp..

**A.4.2 Proposition**
In the situation of Definition A.4.1, the function $A : \mathcal{E} \to \mathcal{F}$ is continuous.

*Proof.*
Let $e_0 \in \mathcal{E}$ and let $\phi : \mathcal{E}_{|U} \to U \times B$ and $\psi : \mathcal{F}_{|V} \to V \times C$ be trivializing maps according to (c) of Definition A.4.1 with $\xi := p(e_0)$. We note

that it suffices to show that $A_{|\mathcal{E}_{|U}}$ is continuous. For all $e \in \mathcal{E}_{|U}$ $A(e) = \psi^{-1}\big(p(e), \psi_{p(e)} A_{p(e)} (\phi_{p(e)})^{-1} \phi_{p(e)} e\big) = \psi^{-1}\big(p(e), \mathcal{A}_{p(e)} \phi_{p(e)} e\big)$. By Fact 1.5.1 $\mathcal{E}_{|U} \ni e \mapsto \mathcal{A}_{p(e)} \in \mathscr{L}(B, C)$ is continuous and by [Que01] Satz 3.10 $\mathcal{E}_{|U} \ni e \mapsto \phi_{p(e)} e \in B$ is continuous. Therefore $\mathcal{E}_{|U} \ni e \mapsto \mathcal{A}_{p(e)} \phi_{p(e)} e \in B$ is continuous. This yields $[e \mapsto A(e)] \in C(\mathcal{E}_{|U}, \mathcal{F}_V)$ and thus $[e \mapsto A(e)] \in C(\mathcal{E}_{|U}, \mathcal{F})$. □

### A.4.3 Proposition

In the situation of Definition A.4.1, condition (c) is equivalent to each of the following conditions.

($c_1$) There are $\{\phi^{(\lambda)} : \mathcal{E}_{|U_\lambda} \to U_\lambda \times B_\lambda\}_{\lambda \in \Lambda} \in \langle \mathcal{E} \overset{p}{\succ} \Omega \rangle$ and $\{\psi^{(\kappa)} : \mathcal{F}_{|V_\kappa} \to V_\kappa \times C_\kappa\}_{\kappa \in K} \in \langle \mathcal{F} \overset{q}{\succ} \Omega \rangle$, such that the trivialized induced maps

$$\mathcal{A}_x^{(\lambda,\kappa)} : B_\lambda \overset{(\phi_x^{(\lambda)})^{-1}}{\longrightarrow} \mathcal{E}_x \overset{A_x}{\longrightarrow} \mathcal{F}_x \overset{\psi_x^{(\kappa)}}{\longrightarrow} C_\kappa$$

are analytically depending on $x \in U_\lambda \cap V_\kappa$ whenever $U_\lambda \cap V_\kappa \neq \emptyset$, i. e. $[x \mapsto \mathcal{A}_x^{(\lambda,\kappa)}] \in A(U_\lambda \cap V_\kappa, \mathscr{L}(B_\lambda, C_\kappa))$.

($c_2$) There are $\{\phi^{(\lambda)} : \mathcal{E}_{|U_\lambda} \to U_\lambda \times B_\lambda\}_{\lambda \in \Lambda} \in \langle \mathcal{E} \overset{p}{\succ} \Omega \rangle$ and $\{\psi^{(\lambda)} : \mathcal{F}_{|U_\lambda} \to U_\lambda \times C_\lambda\}_{\lambda \in \Lambda} \in \langle \mathcal{F} \overset{q}{\succ} \Omega \rangle$ (with the same associated trivializing cover $\{U_\lambda\}_{\lambda \in \Lambda}$ of $\Omega$) such that the trivialized induced maps

$$\mathcal{A}_x^{(\lambda)} : B_\lambda \overset{(\phi_x^{(\lambda)})^{-1}}{\longrightarrow} \mathcal{E}_x \overset{A_x}{\longrightarrow} \mathcal{F}_x \overset{\psi_x^{(\lambda)}}{\longrightarrow} C_\lambda$$

are analytically depending on $x \in U_\lambda$, i. e. $[x \mapsto \mathcal{A}_x^{(\lambda)}] \in A(U_\lambda, \mathscr{L}(B_\lambda, C_\lambda))$ for all $\lambda \in \Lambda$.

($c_3$) *For all* $\{\phi^{(\lambda)} : \mathcal{E}_{|U_\lambda} \to U_\lambda \times B_\lambda\}_{\lambda \in \Lambda} \in \langle \mathcal{E} \overset{p}{\succ} \Omega \rangle$ and $\{\psi^{(\kappa)} : \mathcal{F}_{|V_\kappa} \to V_\kappa \times C_\kappa\}_{\kappa \in K} \in \langle \mathcal{F} \overset{q}{\succ} \Omega \rangle$ the trivialized induced maps

$$\mathcal{A}_x^{(\lambda,\kappa)} : B_\lambda \overset{(\phi_x^{(\lambda)})^{-1}}{\longrightarrow} \mathcal{E}_x \overset{A_x}{\longrightarrow} \mathcal{F}_x \overset{\psi_x^{(\kappa)}}{\longrightarrow} C_\kappa$$

are analytically depending on $x \in U_\lambda \cap V_\kappa$ whenever $U_\lambda \cap V_\kappa \neq \emptyset$, i. e. $[x \mapsto \mathcal{A}_x^{(\lambda,\kappa)}] \in A(U_\lambda \cap V_\kappa, \mathscr{L}(B_\lambda, C_\kappa))$.

*Proof.*
"(c)⇒($c_1$)": For each $\xi \in \Omega$ we denote by $\phi^{(\xi)} : \mathcal{E}_{|U_\xi} \to U_\xi \times B_\xi$ and $\psi^{(\xi)} : \mathcal{F}_{|V_\xi} \to V_\xi \times C_\xi$ the trivializing maps given by (c). We will show that $\{\phi^{(\xi)}\}_{\xi \in \Omega}$

and $\{\psi^{(\xi)}\}_{\xi\in\Omega}$ are trivializations for which ($c_1$) holds: Clearly, $\{U_\xi\}_{\xi\in\Omega}$ is an open cover of $\Omega$ and for all $\xi \in \Omega$ $\phi^{(\xi)}$ is a homeomorphism for which conditions (a) and (b) of Definition A.1.1 hold. Finally, if $U_\xi \cap U_\zeta \neq \emptyset$ for some $\xi, \zeta \in \Omega$ then the transition map $U_\xi \cap U_\zeta \ni x \mapsto \Phi_x^{(\xi,\zeta)} \in \mathscr{L}(B_\xi, B_\zeta)$ fulfills condition (c) of Definition A.1.1, since by definition trivializing maps are part of equivalent trivializing covers. Thus, $\{\psi^{(\xi)}\}_{\xi\in\Omega}$ is a representant of $\langle \mathcal{E} \overset{p}{\succ} \Omega\rangle$. Analogously, $\{\psi^{(\xi)}\}_{\xi\in\Omega}$ is a representant of $\langle \mathcal{F} \overset{q}{\succ} \Omega\rangle$. Now, let $U_\eta \cap V_\zeta \neq \emptyset$ for some $\eta, \zeta \in \Omega$. In order to prove ($c_1$) it suffices to show that $[x \mapsto \mathcal{A}_x^{(\eta,\zeta)}] \in A(U_\eta \cap V_\zeta, \mathscr{L}(B_\eta, C_\zeta))$. To this end, let $\xi \in U_\eta \cap V_\zeta$. We note that it suffices to show that there exist some $W \overset{\circ}{\subset} U_\eta \cap V_\zeta$ with $\xi \in W$ such that $[x \mapsto \mathcal{A}_x^{(\eta,\zeta)}] \in A(W, \mathscr{L}(B_\eta, C_\zeta))$. Let $W := U_\eta \cap V_\zeta \cap U_\xi \cap V_\xi$. Clearly, $\xi \in W \overset{\circ}{\subset} U_\eta \cap V_\zeta$ and for each $x \in W$ the map $B_\eta \xrightarrow{(\phi_x^{(\eta)})^{-1}} \mathcal{E}_x \xrightarrow{A_x} \mathcal{F}_x \xrightarrow{\psi_x^{(\zeta)}} C_\zeta$ can be written as

$$B_\eta \xrightarrow{(\phi_x^{(\eta)})^{-1}} \underbrace{\mathcal{E}_x \xrightarrow{\phi_x^{(\xi)}} B_\xi \xrightarrow{(\phi_x^{(\xi)})^{-1}} \mathcal{E}_x}_{\mathrm{Id}} \xrightarrow{A_x} \underbrace{\mathcal{F}_x \xrightarrow{\psi_x^{(\xi)}} C_\xi \xrightarrow{(\psi_x^{(\xi)})^{-1}} \mathcal{F}_x}_{\mathrm{Id}} \xrightarrow{\psi_x^{(\zeta)}} C_\zeta$$

or, equivalently by associativity,

$$\underbrace{B_\eta \xrightarrow{(\phi_x^{(\eta)})^{-1}} \mathcal{E}_x \xrightarrow{\phi_x^{(\xi)}} B_\xi}_{\phi_x^{(\xi)}\circ(\phi_x^{(\eta)})^{-1}} \underbrace{\xrightarrow{(\phi_x^{(\xi)})^{-1}} \mathcal{E}_x \xrightarrow{A_x} \mathcal{F}_x \xrightarrow{\psi_x^{(\xi)}} C_\xi}_{\psi_x^{(\xi)}\circ A_x\circ(\phi_x^{(\xi)})^{-1}} \underbrace{\xrightarrow{(\psi_x^{(\xi)})^{-1}} \mathcal{F}_x \xrightarrow{\psi_x^{(\zeta)}} C_\zeta}_{\psi_x^{(\zeta)}\circ(\psi_x^{(\xi)})^{-1}}$$

as a "product" of three operator-valued functions and thus by Fact 1.5.10 $[x \mapsto \mathcal{A}_x^{(\eta,\zeta)}] \in A(W, \mathscr{L}(B_\eta, C_\zeta))$.

"($c_1$)⇒($c_2$)": Using the trivializations constructed in the proof of Proposition A.1.10 ($c_2$) directly follows from the assumption.

"($c_2$)⇒($c_3$)": We assume that $\{\phi^{(\lambda)} : \mathcal{E}_{|U_\lambda} \to U_\lambda \times B_\lambda\}_{\lambda\in\Lambda}$ and $\{\psi^{(\lambda)} : \mathcal{F}_{|U_\lambda} \to U_\lambda \times C_\lambda\}_{\lambda\in\Lambda}$ are the trivializations for $\langle \mathcal{E} \overset{p}{\succ} \Omega\rangle$ and $\langle \mathcal{F} \overset{q}{\succ} \Omega\rangle$, resp., given by ($c_2$). Let $\phi : \mathcal{E}_{|U} \to U \times B$ and $\psi : \mathcal{F}_{|V} \to V \times C$ be arbitrary[7] trivializing maps for $\langle \mathcal{E} \overset{p}{\succ} \Omega\rangle$ and $\langle \mathcal{F} \overset{q}{\succ} \Omega\rangle$, resp., with $U \cap V \neq \emptyset$. We note, that it suffices to show $[x \mapsto \psi_x \circ A_x \circ (\phi_x)^{-1}] \in A(U \cap V, \mathscr{L}(B, C))$. In order to prove this, again, it suffices, that for any $\xi \in U \cap V$, there exists $W \overset{\circ}{\subset} U \cap V$ with $\xi \in W$ such that $[x \mapsto \psi_x \circ A_x \circ (\phi_x)^{-1}] \in A(W, \mathscr{L}(B, C))$. Therefore, let $\xi \in U \cap V$.

[7] We remark that we don't assume $\phi \in \{\phi^{(\lambda)}\}_{\lambda\in\Lambda}$ or $\psi \in \{\psi^{(\lambda)}\}_{\lambda\in\Lambda}$.

There exists $\lambda \in \Lambda$ such that $\xi \in U_\lambda$. Then $W := U_\lambda \cap U \cap V \overset{\circ}{\subset} U \cap V$ and $\xi \in W$. Analogously to step "(c)$\Rightarrow$(c$_1$)", for each $x \in W$ the map $B \xrightarrow{(\phi_x)^{-1}} \mathcal{E}_x \xrightarrow{A_x} \mathcal{F}_x \xrightarrow{\psi_x} C$ can be written as

$$B \underbrace{\xrightarrow{(\phi_x)^{-1}} \mathcal{E}_x \xrightarrow{\phi_x^{(\lambda)}}}_{\phi_x^{(\lambda)} \circ (\phi_x)^{-1}} B_\lambda \underbrace{\xrightarrow{(\phi_x^{(\lambda)})^{-1}} \mathcal{E}_x \xrightarrow{A_x} \mathcal{F}_x \xrightarrow{\psi_x^{(\lambda)}}}_{\psi_x^{(\lambda)} \circ A_x \circ (\phi_x^{(\lambda)})^{-1}} C_\lambda \underbrace{\xrightarrow{(\psi_x^{(\lambda)})^{-1}} \mathcal{F}_x \xrightarrow{\psi_x}}_{\psi_x \circ (\psi_x^{(\lambda)})^{-1}} C$$

and thus again by Fact 1.5.10 $[x \mapsto \psi_x \circ A_x \circ (\phi_x)^{-1}] \in A(W, \mathscr{L}(B, C))$. This proves (c$_3$).

"(c$_3$)$\Rightarrow$(c)": For each $\xi \in \Omega$ let $\phi : \mathcal{E}_{|U} \to U \times B$, $\psi : \mathcal{F}_{|V} \to V \times C$ be trivializing maps with $\xi \in U$, $\xi \in V$ (such maps obviously exit, since every bundle has a representant by definition). By Proposition A.1.9 and the assumption, $\phi_{|U \cap V}$ and $\psi_{|U \cap V}$ are trivializing maps for which (c) holds. □

**A.4.4 Definition** (*Isomorphism*)
Let $A : \langle \mathcal{E} \overset{p}{\succ} \Omega \rangle \to \langle \mathcal{F} \overset{q}{\succ} \Omega \rangle$ be a bijective homomorphism of bundles. Furthermore, assume that $A$ is bijective and that $A^{-1} : \mathcal{F} \to \mathcal{E}$ is a homomorphism, too.
Then $A$ is called an *isomorphism (of bundles)* and $\langle \mathcal{E} \overset{p}{\succ} \Omega \rangle$ and $\langle \mathcal{F} \overset{q}{\succ} \Omega \rangle$ are called *isomorphic.*

**A.4.5 Proposition**
Let $A : \langle \mathcal{E} \overset{p}{\succ} \Omega \rangle \to \langle \mathcal{F} \overset{q}{\succ} \Omega \rangle$ be a homomorphism of bundles over $\Omega$.
Then the following are all equivalent.

(a) $A$ is an isomorphism.

(b) $A$ is bijective.

(c) For each $x \in \Omega$ $A_x$ is an isomorphism.

*Proof.*
"(a)$\Rightarrow$(b)" and "(b)$\Rightarrow$(c)" are clear. In order to show "(c)$\Rightarrow$(a)" we assume that (b) holds. Clearly, bijectivity of $A_x$ for each $x \in \Omega$ in combination with condition (a) of Definition A.4.1 yields bijectivity of $A$. Therefore it remains to show that $A^{-1} : \mathcal{F} \to \mathcal{E}$ is a homomorphism. Bijectivity of $A$ implies conditions (a) and (b) of Definition A.4.1 for $A^{-1}$. By Proposition A.4.3 (c$_2$) there are $\{\phi^{(\lambda)} : \mathcal{E}_{|U_\lambda} \to U_\lambda \times B_\lambda\}_{\lambda \in \Lambda} \in \langle \mathcal{E} \overset{p}{\succ} \Omega \rangle$ and $\{\psi^{(\lambda)} : \mathcal{F}_{|U_\lambda} \to U_\lambda \times C_\lambda\}_{\lambda \in \Lambda} \in \langle \mathcal{F} \overset{q}{\succ} \Omega \rangle$ such that $[x \mapsto \mathcal{A}_x^{(\lambda)}] \in A(U_\lambda, \mathscr{L}(B_\lambda, C_\lambda))$ for all $\lambda \in \Lambda$, where $\mathcal{A}_x^{(\lambda)}$

denotes the trivialized induced maps of the bundle homomorphism $A$. Then the trivialized induced maps associated with $A^{-1}$ are given by $\tilde{\mathcal{A}}_x^{(\lambda)} : C_\lambda \xrightarrow{(\psi_x^{(\lambda)})^{-1}} \mathcal{F}_x \xrightarrow{(A_x)^{-1}} \mathcal{E}_x \xrightarrow{\phi_x^{(\lambda)}} B_\lambda$ and thus coincides with $(\mathcal{A}_x^{(\lambda)})^{-1}$. Hence by [Cha85] Theorems 7.17, 5.9 and 14.13 $[x \mapsto \tilde{\mathcal{A}}_x^{(\lambda)}] \in A(U_\lambda, \mathscr{L}(C_\lambda, B_\lambda))$ for all $\lambda \in \Lambda$. Therefore Proposition A.4.3 yields condition (c) of Definition A.4.1 for $A^{-1}$. □

**A.4.6 Proposition** (*Compatibility w. r. t. Equivalence*)
Let $A : \langle \mathcal{E} \overset{p}{\succ} \Omega \rangle \to \langle \mathcal{F} \overset{q}{\succ} \Omega \rangle$ be an isomorphism of bundles and let $\{\phi^{(\lambda)}\}_{\lambda \in \Lambda}$ be a representant of $\langle \mathcal{E} \overset{p}{\succ} \Omega \rangle$.
Then $\{\phi^{(\lambda)} \circ A^{-1}\}_{\lambda \in \Lambda}$ is a representant of $\langle \mathcal{F} \overset{q}{\succ} \Omega \rangle$.

*Proof.*
For each $\lambda \in \Lambda$ we denote by $U_\lambda$ and $C_\lambda$ the open set and the Banach space associated with $\phi^{(\lambda)}$, i. e. $\phi^{(\lambda)}$ has the form $\phi^{(\lambda)} : \mathcal{E}_{|U_\lambda} \to U_\lambda \times B_\lambda$. By definition of $A^{-1}$ and $\phi^{(\lambda)}$ for each $f \in \mathcal{F}_{|U_\lambda}$ $q(f) = p(A^{-1}f) = (\nu \circ \phi^{(\lambda)})(A^{-1}f) = \nu(\phi^{(\lambda)} \circ A^{-1})(f))$, where $\nu$ is the natural projection (associated with both $\{\phi^{(\lambda)}\}_{\lambda \in \Lambda}$ and $\{\phi^{(\lambda)} \circ A^{-1}\}_{\lambda \in \Lambda}$), and thus condition (a) of Definition A.1.1 holds for $\phi^{(\lambda)} \circ A^{-1}$. Furthermore, on each fiber $\mathcal{F}_x$ (where $x \in U_\lambda$) the map induced by $\phi^{(\lambda)} \circ A^{-1}$ is given by $(\phi^{(\lambda)} \circ A^{-1})_x : \mathcal{F}_x \xrightarrow{(A^{-1})_x} \mathcal{E}_x \xrightarrow{\phi_x^{(\lambda)}} B_\lambda$ and thus is an isomorphism. Therefore condition (b) holds. Additionally, for each $\kappa \in \Lambda$ with $U_\lambda \cap U_\kappa \neq \emptyset$ then the transition function from $B_\lambda$ to $B_\kappa$ w. r. t. $\{\phi^{(\lambda)} \circ A^{-1}\}_{\lambda \in \Lambda}$ is given by $B_\lambda \xrightarrow{(\phi_x^{(\lambda)})^{-1}} \mathcal{E}_x \xrightarrow{((A^{-1})_x)^{-1}} \mathcal{F}_x \xrightarrow{(A^{-1})_x} \mathcal{E}_x \xrightarrow{\phi_x^{(\kappa)}} B_\kappa$. Since $((A^{-1})_x)^{-1} = A_x$ conditions (c) holds. Thus $\{\phi^{(\lambda)} \circ A^{-1}\}_{\lambda \in \Lambda}$ is a trivialization for $q$. Finally, if $\psi : V \to V \times C$ is trivializing map for $\langle \mathcal{F} \overset{q}{\succ} \Omega \rangle$ with $U_\lambda \cap V \neq \emptyset$ then the transition function from $B_\lambda$ to $C$ is given by $B_\lambda \xrightarrow{(\phi_x^{(\lambda)})^{-1}} \mathcal{E}_x \xrightarrow{A_x} \mathcal{F}_x \xrightarrow{\psi_x} C$ and thus is analytically depending on $x \in U_\lambda \cap V$ by Proposition A.4.3 ($c_3$). Together with Remark A.1.2 this yields that $\{\phi^{(\lambda)} \circ A^{-1}\}_{\lambda \in \Lambda}$ is a representant of $\langle \mathcal{F} \overset{q}{\succ} \Omega \rangle$. □

**A.4.7 Corollary**
Let $A : \langle \mathcal{E} \overset{p}{\succ} \Omega \rangle \to \langle \mathcal{F} \overset{q}{\succ} \Omega \rangle$ be an isomorphism of bundles.
Then every representant of $\langle \mathcal{F} \overset{q}{\succ} \Omega \rangle$ is of the form $\{\phi^{(\lambda)} \circ A^{-1}\}_{\lambda \in \Lambda}$ where $\{\phi^{(\lambda)}\}_{\lambda \in \Lambda}$ is a representant of $\langle \mathcal{E} \overset{p}{\succ} \Omega \rangle$.

*Proof.*
Let $\{\psi^{(\lambda)}\}_{\lambda\in\Lambda}$ be a representant of $\langle\mathcal{F}\overset{q}{\succ}\Omega\rangle$. Then by Proposition A.4.6 $\{\psi^{(\lambda)}\circ A\}_{\lambda\in\Lambda}$ is a representant of $\langle\mathcal{E}\overset{q}{\succ}\Omega\rangle$. The result follows with $\phi^{(\lambda)} := \psi^{(\lambda)}\circ A$ (where $\lambda\in\Lambda$). □

**A.4.8 Example**
Let $\{\phi^{(\lambda)} : \mathcal{E}_{|U_\lambda} \to U_\lambda \times B_\lambda\}_{\lambda\in\Lambda}$ be a trivialization for a bundle projection $p : \mathcal{E} \to \Omega$. For each $x \in \Omega$ let $\|\cdot\|_x$ be a norm on the Banach space $\mathcal{E}_x$ that is equivalent to the "original" one. Denote by $\tilde{\mathcal{E}} = \bigcup_{x\in\Omega}(\mathcal{E}_x, \|\cdot\|_x)$ the total space "associated" with the norms $\|\cdot\|_x$. Then $\{\phi^{(\lambda)}\}_{\lambda\in\Lambda}$ is a trivialization for $p : \tilde{\mathcal{E}} \to \Omega$ since obviously conditions (a), (b) and (c) of Definition A.1.1 hold. Thus its equivalence class $\langle\tilde{\mathcal{E}}\overset{p}{\succ}\Omega\rangle$ is a bundle. Then $\mathrm{Id} : \mathcal{E} \to \tilde{\mathcal{E}}$ is an isomorphism of bundles: Obviously, Id is bijective and condition (a) of Definition A.4.1 holds. Furthermore, clearly the induced operator on the fibers $\mathrm{Id}_x$ is linear and by equivalence of the norms $\mathrm{Id}_x \in \mathscr{L}(\mathcal{E}_x, \tilde{\mathcal{E}}_x)$. Thus condition (b) holds. Finally, let $x \in \Omega$ and $\lambda \in \Lambda$ such that $x \in U_\lambda$. The transition function $\mathcal{I}_x : B_\lambda \xrightarrow{(\phi_x^{(\lambda)})^{-1}} \mathcal{E}_x \xrightarrow{\mathrm{Id}_x} \tilde{\mathcal{E}}_x \xrightarrow{\tilde{\phi}_x^{(\lambda)}} B_\lambda$ coincides with $\mathrm{Id}_{B_\lambda}$ and thus is analytically depending on $x \in U_\lambda$. Therefore condition (c) holds. △

### Spectrum and Cospectrum

**A.4.9 Definition**
Let $A : \langle\mathcal{E}\overset{p}{\succ}\Omega\rangle \to \langle\mathcal{F}\overset{q}{\succ}\Omega\rangle$ be a homomorphism of bundles.

$\mathrm{S}(A) := \{\, x \in \Omega : \operatorname{Ker} A_x \neq \{0\}\,\}$ is called the *spectrum (of A)*.

$\mathrm{CS}(A) := \{\, x \in \Omega : \operatorname{Coker} A_x \neq \{0\}\,\}$ is called the *cospectrum (of A)*.

## A.5 Trivial Bundles

**A.5.1 Definition** (*Trivial Bundle*)
A bundle, that is isomorphic to the bundle constructed in Example A.1.8, is called *trivial bundle*.

**A.5.2 Proposition** (*A Characterization of Trivial Bundles*)
A bundle $\langle\mathcal{E}\overset{p}{\succ}\Omega\rangle$ is trivial iff it has a representant of the form $\{\phi : \mathcal{E}_{|\Omega} \to$

$\Omega \times B\}$.

*Proof.*
By definition, the bundle constructed in Example A.1.8 has such a representant. Thus Proposition A.4.6 yields the direction "$\Longrightarrow$". Conversely for "$\Longleftarrow$", let $\{\phi : \mathcal{E}_{|\Omega} \to \Omega \times B\}$ be a representant of $\langle \mathcal{E} \overset{p}{\succ} \Omega\rangle$. Denote by $\langle \Omega \times B \overset{\nu}{\succ} \Omega\rangle$ the bundle constructed in Example A.1.8 (with fibers the given Banach space $B$). Then $\phi : \langle \mathcal{E} \overset{p}{\succ} \Omega\rangle \to \langle \Omega \times B \overset{\nu}{\succ} \Omega\rangle$ is an isomorphism: By definition, $\phi$ is bijective and conditions (a) and (b) of Definition A.1.1 yield conditions (a) and (b) of Definition A.4.1. Finally, let $x \in \Omega$. Then the trivialized induced map $B \overset{(\phi_x)^{-1}}{\longrightarrow} \mathcal{E}_x \overset{\phi_x}{\longrightarrow} \{x\} \times B \overset{\mathrm{Id}_x}{\longrightarrow} B$ (where, by abuse of notation, $(\phi_x)^{-1}$ denotes the induced isomorphism in the sense of Definition A.1.1 (b) and $\phi_x$ denotes the induced operator in the sense of Definition A.4.1 (b))] coincides with $\mathrm{Id}_B$ and thus is analytically depending on $x \in \Omega$. Therefore condition (c) of Definition A.4.1 holds. □

## A.6 Subbundles

### A.6.1 Definition
Let $\langle \mathcal{E} \overset{p}{\succ} \Omega\rangle$ and $\langle \mathcal{F} \overset{q}{\succ} \Omega\rangle$ be bundles over the same base space $\Omega$.
$\langle \mathcal{E} \overset{p}{\succ} \Omega\rangle$ is called a *subbundle (of* $\langle \mathcal{F} \overset{q}{\succ} \Omega\rangle$*)* if

(a) $\mathcal{E} \subset \mathcal{F}$ (and the topology of $\mathcal{E}$ is the induced one from $\mathcal{F}$ and for each $x \in \Omega$ the Banach space $\mathcal{E}_x$ is a subspace of the Banach space $\mathcal{F}_x$) and

(b) the inclusion map $I : \mathcal{E} \hookrightarrow \mathcal{F}$ is an homomorphism from $\langle \mathcal{E} \overset{p}{\succ} \Omega\rangle$ to $\langle \mathcal{F} \overset{q}{\succ} \Omega\rangle$.

### A.6.2 Remark
By Proposition A.4.3 and Proposition A.1.10, in the situation of Definition A.6.1 condition (b) is equivalent to the following condition.

($b_1$) $p = q_{|\mathcal{E}}$ and for all $\xi \in \Omega$ there are trivializing maps $\phi : \mathcal{E}_{|U} \to U \times B$ for $\langle \mathcal{E} \overset{p}{\succ} \Omega\rangle$ and $\psi : \mathcal{F}_{|U} \to U \times C$ for $\langle \mathcal{F} \overset{q}{\succ} \Omega\rangle$, resp., with $\xi \in U$ such that the map

$$\mathcal{I}_x : B \overset{(\phi_x)^{-1}}{\longrightarrow} \mathcal{E}_x \hookrightarrow \mathcal{F}_x \overset{\psi_x}{\longrightarrow} C$$

is analytically depending on $x \in U$, i.e. $[x \mapsto \mathcal{I}_x] \in A(U, \mathscr{L}(B, C))$, where for each $x \in U$ $\mathcal{E}_x \hookrightarrow \mathcal{F}_x$ denotes the embedding by inclusion as a subspace.

We refer to Proposition A.4.3 and Definition A.4.1 for further equivalent formulations.

**A.6.3 Remark**
In the situation of Definition A.6.1 $\mathcal{E}_x = \mathcal{F}_x \cap \mathcal{E}$ for each $x \in \Omega$ and thus $\mathcal{E}_x$ is a closed linear subspace of $\mathcal{F}_x$ since all $\mathcal{E}_x$ are complete.

**A.6.4 Remark**
Clearly, every bundle is a subbundle of itself.

**A.6.5 Remark**
Let $\langle \mathcal{E} \overset{p}{\succ} \Omega \rangle$ be a subbundle of $\langle \mathcal{F} \overset{q}{\succ} \Omega \rangle$ and $s \in \Gamma(O, \langle \mathcal{E} \overset{p}{\succ} \Omega \rangle)$ where $\emptyset \neq O \overset{\circ}{\subset} \Omega$.
For each $\xi \in O$ by Proposition A.1.10 there are trivializing maps $\phi : \mathcal{E}_{|U} \to U \times B$ for $\langle \mathcal{E} \overset{p}{\succ} \Omega \rangle$ and $\psi : \mathcal{F}_{|U} \to U \times C$ for $\langle \mathcal{F} \overset{q}{\succ} \Omega \rangle$, resp., with $\xi \in U \subset O$. Then by Remark A.6.2, Proposition A.2.4 and Fact 1.5.10 $q \circ s = p \circ s = \mathrm{Id}_O$ and $\big[x \mapsto \psi_x\big(s(x)\big)\big] = \big[x \mapsto \psi_x(\phi_x)^{-1}\phi_x\big(s(x)\big)\big] \in A(O, C)$. Hence, again by Proposition A.2.4 $s \in \Gamma(O, \langle \mathcal{F} \overset{q}{\succ} \Omega \rangle)$.

## A.7 Induced Homomorphisms on Sections

**A.7.1 Proposition and Definition**
Let $A : \langle \mathcal{E} \overset{p}{\succ} \Omega \rangle \to \langle \mathcal{F} \overset{q}{\succ} \Omega \rangle$ be a homomorphism of bundles and $\emptyset \neq O \overset{\circ}{\subset} \Omega$. Then $A_{\Gamma|O}(s) := A \circ s$ for all $s \in \Gamma(O, \langle \mathcal{E} \overset{p}{\succ} \Omega \rangle)$ defines a map $A_{\Gamma|O} \in \mathscr{L}\big(\Gamma(O, \langle \mathcal{E} \overset{p}{\succ} \Omega \rangle), \Gamma(O, \langle \mathcal{F} \overset{p}{\succ} \Omega \rangle)\big)$, called *the induced homomorphism by* $A$. We set $A_\Gamma := A_{\Gamma|\Omega}$.

*Proof.*
Let $s \in \Gamma(O, \langle \mathcal{E} \overset{p}{\succ} \Omega \rangle)$. Then $q \circ \big(A_{\Gamma|O}(s)\big) = q \circ A \circ s = p \circ s = \mathrm{Id}_O$. Furthermore, let $\xi \in O$. By Proposition A.4.3 (2) there are trivializing maps $\phi : \mathcal{E}_{|U} \to U \times B$ for $\langle \mathcal{E} \overset{p}{\succ} \Omega \rangle$ and $\psi : \mathcal{F}_{|V} \to V \times C$ for $\langle \mathcal{F} \overset{q}{\succ} \Omega \rangle$, resp., such that Definition A.4.1 (c) holds. Then for all $x \in U \cap O$ $\psi_x\big(\big(A_{\Gamma|O}(s)\big)(x)\big) = \psi_x\big(A\big(s(x)\big)\big) = \psi_x\big(A_x\big((\phi_x^{-1})\phi_x s(x)\big)\big) = (\psi_x \circ A_x \circ \phi_x^{-1})\big(\phi_x s(x)\big)$. By the assumption $[x \mapsto \psi_x \circ A_x \circ \phi_x^{-1}] \in A(U, \mathscr{L}(B, C))$ and $\big[x \mapsto \phi_x\big(s(x)\big)\big] \in$

$A(U \cap O, B)$. Thus by Fact 1.5.10 $\big[x \mapsto \psi_x\big(\big(A_{\Gamma|O}(s)\big)(x)\big)\big] \in A(U \cap O, C)$. This yields $A_{\Gamma|O}(s) \in \langle \mathcal{F} \overset{p}{\succ} \Omega \rangle$. Clearly, $A_{\Gamma|O}$ is linear. In order to prove continuity it suffices to show that $(A_{\Gamma|O})^{-1}\big(B_{\Gamma(O,\langle \mathcal{F} \overset{p}{\succ} \Omega \rangle)}(K, \mathcal{O})\big)$ is open for every $K \subset\subset O$ and $\mathcal{O} \overset{\circ}{\subset} \mathcal{F}$. By Proposition A.4.2 $A^{-1}(\mathcal{O}) \overset{\circ}{\subset} \mathcal{E}$ and thus $(A_{\Gamma|O})^{-1}(B_{\Gamma(O,\langle \mathcal{F} \overset{p}{\succ} \Omega \rangle)}(K, \mathcal{O})) = B_{\Gamma(O,\langle \mathcal{E} \overset{p}{\succ} \Omega \rangle)}(K, A^{-1}(\mathcal{O})) \overset{\circ}{\subset} \Gamma(O, \langle \mathcal{E} \overset{p}{\succ} \Omega \rangle)$ for every $K \subset\subset O$ and $\mathcal{O} \overset{\circ}{\subset} \mathcal{F}$. This finishes the proof. □

## A.8 Fredholm Homomorphisms

### A.8.1 Definition

A homomorphism $A : \langle \mathcal{E} \overset{p}{\succ} \Omega \rangle \to \langle \mathcal{F} \overset{q}{\succ} \Omega \rangle$ of bundles over the space $\Omega$ is called *Fredholm homomorphism* if for each $x \in \Omega$ the induced operator $A_x \in \mathscr{L}(\mathcal{E}_x, \mathcal{F}_x)$ is a Fredholm operator.

### A.8.2 Proposition

Let $A : \langle \mathcal{E} \overset{p}{\succ} \Omega \rangle \to \langle \mathcal{F} \overset{q}{\succ} \Omega \rangle$ be a homomorphism of bundles.
Then the following are all equivalent.

(1) $A$ is a Fredholm homomorphism.

(2) There are trivializations $\{\phi^{(\lambda)} : \mathcal{E}_{|U_\lambda} \to U_\lambda \times B_\lambda\}_{\lambda \in \Lambda}$ and $\{\psi^{(\kappa)} : \mathcal{F}_{|V_\kappa} \to V_\kappa \times C_\kappa\}_{\kappa \in K}$ for $\langle \mathcal{E} \overset{p}{\succ} \Omega \rangle$ and $\langle \mathcal{F} \overset{q}{\succ} \Omega \rangle$, resp., such that the trivialized induced maps

$$\mathcal{A}_x^{(\lambda,\kappa)} : B_\lambda \overset{(\phi_x^{(\lambda)})^{-1}}{\longrightarrow} \mathcal{E}_x \overset{A_x}{\longrightarrow} \mathcal{F}_x \overset{\psi_x^{(\kappa)}}{\longrightarrow} C_\kappa$$

are Fredholm operators whenever $x \in U_\lambda \cap V_\kappa \neq \emptyset$.

(3) There are trivializations $\{\phi^{(\lambda)} : \mathcal{E}_{|U_\lambda} \to U_\lambda \times B_\lambda\}_{\lambda \in \Lambda}$ and $\{\psi^{(\lambda)} : \mathcal{F}_{|U_\lambda} \to U_\lambda \times C_\lambda\}_{\lambda \in \Lambda}$ for $\langle \mathcal{E} \overset{p}{\succ} \Omega \rangle$ and $\langle \mathcal{F} \overset{q}{\succ} \Omega \rangle$, resp., (with the same open cover $\{U_\lambda\}_{\lambda \in \Lambda}$ of $\Omega$) such that the trivialized induced maps

$$\mathcal{A}_x^{(\lambda)} : B_\lambda \overset{(\phi_x^{(\lambda)})^{-1}}{\longrightarrow} \mathcal{E}_x \overset{A_x}{\longrightarrow} \mathcal{F}_x \overset{\psi_x^{(\lambda)}}{\longrightarrow} C_\lambda$$

are Fredholm operators for all $x \in U_\lambda$ for all $\lambda \in \Lambda$.

(4) *For all* trivializations $\{\phi^{(\lambda)} : \mathcal{E}_{|U_\lambda} \to U_\lambda \times B_\lambda\}_{\lambda \in \Lambda}$ and $\{\psi^{(\kappa)} : \mathcal{F}_{|V_\kappa} \to$

$V_\kappa \times C_\kappa\}_{\kappa\in K}$ for $\langle \mathcal{E} \overset{p}{\succ} \Omega\rangle$ and $\langle \mathcal{F} \overset{q}{\succ} \Omega\rangle$, resp., the trivialized induced maps

$$\mathcal{A}_x^{(\lambda,\kappa)} : B_\lambda \overset{(\phi_x^{(\lambda)})^{-1}}{\longrightarrow} \mathcal{E}_x \overset{A_x}{\longrightarrow} \mathcal{F}_x \overset{\psi_x^{(\kappa)}}{\longrightarrow} C_\kappa$$

are Fredholm operators whenever $x \in U_\lambda \cap V_\kappa \neq \emptyset$.

*Proof.*
"(1)⇒(2)": Let $\{\phi^{(\lambda)}\}_{\lambda\in\Lambda}$ and $\{\psi^{(\kappa)}\}_{\kappa\in K}$ be the trivializations from A.4.3 ($c_1$). Since the composition of isomorphisms with Fredholm operators are Fredholm operators (2) follows from the assumption.

"(2)⇒(3)": This follows from the same argument as the step "($c_1$)⇒($c_2$)" in the proof of Proposition A.4.3.

"(3)⇒(4)": Using the construction and its notation from the step "($c_2$)⇒($c_3$)" in the proof of Proposition A.4.3 we obtain the operator composition

$$B \overset{(\phi_x)^{-1}}{\longrightarrow} \mathcal{E}_x \overset{\phi_x^{(\lambda)}}{\longrightarrow} B_\lambda \overset{(\phi_x^{(\lambda)})^{-1}}{\longrightarrow} \mathcal{E}_x \overset{A_x}{\longrightarrow} \mathcal{F}_x \overset{\psi_x^{(\lambda)}}{\longrightarrow} C_\lambda \overset{(\psi_x^{(\lambda)})^{-1}}{\longrightarrow} \mathcal{F}_x \overset{\psi_x}{\longrightarrow} C$$

$$\underbrace{\phi_x^{(\lambda)}\circ(\phi_x)^{-1}}_{B\to B_\lambda} \qquad \underbrace{\psi_x^{(\lambda)}\circ A_x\circ(\phi_x^{(\lambda)})^{-1}}_{B_\lambda\to C_\lambda} \qquad \underbrace{\psi_x\circ(\psi_x^{(\lambda)})^{-1}}_{C_\lambda\to C}$$

which together with the argument used in the step "(1)⇒(2)" proves (4).

"(4)⇒(1)": For each $x \in \Omega$ let $\phi : \mathcal{E}_{|U} \to U\times B$, $\psi : \mathcal{F}_{|V} \to V\times C$ be trivializing maps with $x \in U$, $x \in V$ (such maps obviously exit, since every bundle has a trivialization by definition). Then $A_x = (\psi_x)^{-1} \circ \psi_x \circ A_x \circ (\phi_x)^{-1} \circ \phi_x$ and by the assumption and the argument of step "(1)⇒(2)", (1) follows. □

# Appendix B

# Sheaves

For the convenience of the reader we recall some basic definitions from sheaf theory. We remark that the main intention of this chapter is to introduce and fix notations. We refer to, e. g., [Hör67] Chapter 7, [GR65] Chapter 4 and the monographs [Kul70], [Ten75] and [GR84] where the reader will find proofs that we will omit here.

Furthermore, we state well-known results that we use in this thesis.

We will restrict ourselves to analytic sheaves over open subsets of $\mathbb{C}$, as only those occur during this thesis, and the following notions will be used only in the "customized" version defined below.

During this chapter, let $\emptyset \neq \Omega \overset{\circ}{\subset} \mathbb{C}$.

**B.1.3 Definition and Remark** (*Local Homeomorphism*)
Let $\mathcal{F}$ be a topological space. Then $p : \mathcal{F} \to \Omega$ is called a *local homeomorphism* iff for every $a \in \mathcal{F}$ there are $U \overset{\circ}{\subset} \mathcal{F}$ and $O \overset{\circ}{\subset} \Omega$ such that $a \in U$ and $p_{|U} : U \to O$ is a homeomorphism.
A local homeomorphism is continuous and open (see e. g. [Ten75] Lemma 1.3.5).

**B.1.4 Convention** (*Ring, Module*)
With *ring* we always mean a commutative ring with identity. The underlying ring $R$ of a *module* $M$ ("$R$-module $M$" for short) will always be a ring in the sense above. The underlying ring homomorphism of an *algebra* is always assumed to be of the form $\mathbb{C} \to M$ where $M$ is a ring in the sense above; in particular, we will only deal with $\mathbb{C}$-algebras.

(We refer to e. g. [Bos04] Sections 2.1, 2.9 and 3.3, resp., for precise definitions of the mentioned structures and corresponding homomorphisms. For intuition, we remark that loosely speaking, a ring is a field without division, i. e. there are associative, commutative and "compatible" operations "+", "−" and "·", and corresponding neutral elements 0 and 1. Also, loosely speaking, a module $M$ is a vector space where the field is replaced by a ring $R$, i. e. there are operations "+" and "−" for elements in $M$ a corresponding neutral element 0 and a "compatible" scalar multiplication $\cdot : R \times M \to M$. Furthermore, an algebra is a ring that additionally allows a compatible scalar multiplication with elements of $\mathbb{C}$.)

As an example, the space of analytic functions $A(\Omega)$ is a $\mathbb{C}$-algebra; again loosely speaking that means, that for all $f, g \in A(\Omega)$ and $\alpha \in \mathbb{C}$ there are operations $f + g$, $f - g$, $f \cdot g$, $\alpha f$ and the constant analytic functions $0, \mathbb{1}$ are the corresponding neutral elements. Furthermore, e. g. $A(\Omega, \mathbb{C}^2)$ is an module over $A(\Omega)$.

**B.1.5 Construction** (*The Sheaf of Germs of Analytic Functions*)
For all $z \in \Omega$ and functions $f : \Omega \supset D_f \to \mathbb{C}$ and $g : \Omega \supset D_g \to \mathbb{C}$ we write $f \sim_z g$ iff there is a neighborhood $O$ of $z$, such that $O \subset D_f \cap D_g$, $f, g \in A(O)$ and $f = g$ on $O$.

Then $\sim_z$ is an equivalence relation and the residue class $\gamma_z(f)$ of such $f$ is called the *germ* (of $f$ at $z$).

Furthermore, let $\mathcal{O}_z$ be the set of all residue classes with respect to $\sim_z$ and $\mathcal{O}(\Omega) := \dot{\bigcup}_{z \in \Omega} \mathcal{O}_z$ their disjoint union.

Obviously, if $a \in \mathcal{O}_z$, then there exists a function $f : \Omega \supset D_f \to \mathbb{C}$ with $a = \gamma_z(f)$ and $a(z) := f(z)$ is well-defined.

For each $z \in \Omega$ $\mathcal{O}_z$ is endowed with the ring structure $(\mathcal{O}_z, +_z, \cdot_z)$ induced by the ring structure of the space of analytic functions, i. e. if for each $i = 1, 2$ $a_i \in \mathcal{O}_z$ and $f_i \in A(O_i)$, then the ring operations are given by $a_1 +_z a_2 := \gamma_z(\tilde{f}_1 + \tilde{f}_1)$, $a_1 \cdot_z a_2 := \gamma_z(\tilde{f}_1 \cdot \tilde{f}_1)$, where $\tilde{f}_i$ denotes the restriction of $f_i$ to $O_1 \cap O_2$, and the neutral elements are the residue classes of $0, \mathbb{1} \in A(\Omega)$, resp..

We define a function $p : \mathcal{O}(\Omega) \to \Omega$: If $a \in \mathcal{O}(\Omega)$ then by definition of the disjoint union, there exists exactly one $p(a) \in \Omega$ with $a \in \mathcal{O}_{p(a)}$. $p$ is called *projection*.

We endow $\mathcal{O}(\Omega)$ with the topology that is defined by the base of all sets of the form $\{f \lhd O\} := \{\, \gamma_z(f) : z \in O \,\}$ where $O \overset{\circ}{\subset} \Omega$ and $f \in A(O)$, i. e. a subset of $\mathcal{O}(\Omega)$ is open iff it is the union of sets of this form.

Then $p$ is a local homeomorphism: Indeed, for all $\emptyset \neq O \overset{\circ}{\subset} \Omega$ and $f \in A(O)$ the restriction $p_{|\{f \lhd O\}} : \begin{cases} \{f \lhd O\} & \to & O \\ \gamma_z(f) & \mapsto & z \end{cases}$ is a homeomorphism.

$\mathcal{O}(\Omega)$ is called the *sheaf of germs of analytic functions (over $\Omega$)*, cf. Remark B.1.8.

**B.1.6 Definition** (*Analytic Sheaf*)
Let $\mathcal{F}$ be a topological space. Furthermore, let $p : \mathcal{F} \to \Omega$ be surjective and a local homeomorphism.

For each $z \in \Omega$ $\mathcal{F}_z := p^{-1}(\{z\})$ is called *stalk*. The elements of $\mathcal{F}_z$ are called *germs*.

Furthermore, assume that each stalk $\mathcal{F}_z$ carries the structure of an $\mathcal{O}_z$-module. We set $\mathcal{F} \circ \mathcal{F} = \bigcup_{z \in \Omega} \mathcal{F}_z \times \mathcal{F}_z$ and $\mathcal{O}(\Omega) \circ \mathcal{F} = \bigcup_{z \in \Omega} \mathcal{O}_z \times \mathcal{F}_z$.

We define $+ : \mathcal{F} \circ \mathcal{F} \to \mathcal{F}$ and $\cdot : \mathcal{O}(\Omega) \circ \mathcal{F} \to \mathcal{F}$ by the pointwise operations of the corresponding module. Finally, assume that $+$, $\cdot$ and $X \ni x \mapsto 0_x \in \mathcal{F}$ (where $0_x$ is the neutral element in $\mathcal{F}_x$) are continuous.

Then (by abuse of notation) $\mathcal{F}$ is called an *(analytic) sheaf (over $\Omega$)*.

$p$ is called *projection*.

**B.1.7 Remark**
If $\mathcal{F}$ is a sheaf, then the induced topology (by $\mathcal{F}$) on each stalk $\mathcal{F}_z$ is the discrete topology (i. e. every subset is open).

**B.1.8 Remark**
$\mathcal{O}(\Omega)$ is a sheaf. (For each $z \in \Omega$ $\mathcal{O}_z$ is a module over itself). We also remark that $\Omega \ni z \mapsto 1_z \in \mathcal{O}(\Omega)$ (where $1_z$ is the identity in $\mathcal{O}_z$) is continuous.

**B.1.9 Definition and Fact**
Let $\mathcal{F}$ be a sheaf over $\Omega$ and $\emptyset \neq O \overset{\circ}{\subset} \Omega$. A continuous map $s : O \to \mathcal{F}$ with $p \circ s = \mathrm{Id}$ is called a *section (over $O$)*.

We denote the set of all sections over $O$ by $\Gamma(O, \mathcal{F})$.

The stalkwise ring structure of $\mathcal{O}(\Omega)$ (or, viewed from a different point[1]:

[1]Cf. Remark B.1.10.

the ring structure of $A(\Omega)$) induces a ring structure on $\Gamma(O, \mathcal{O}(\Omega))$ and the stalkwise module structure of $\mathcal{F}$ induces a $\Gamma(O, \mathcal{O}(\Omega))$ module structure on $\Gamma(O, \mathcal{F})$, cf. [Kul70] I.§ 2. Moreover, the $\mathbb{C}$-algebra structure of $A(\Omega)$ induces $\mathbb{C}$-vector space structures on both $\Gamma(O, \mathcal{O}(\Omega))$ and $\Gamma(O, \mathcal{F})$.

**B.1.10 Remark** (*Identification of $A(O)$ with $\Gamma(O, \mathcal{O}(\Omega))$*)
Let $\emptyset \neq O \overset{\circ}{\subset} \Omega$.

For all $f \in A(O)$ we define $\sigma(f) : O \to \mathcal{O}$ by $[\sigma(f)](z) := \gamma_z(f)$. Then $\sigma_f \in \Gamma(O, \mathcal{O}(\Omega))$.

The so-defined map $\sigma : A(O) \to \Gamma(O, \mathcal{O}(\Omega))$ is bijective.

**B.1.11 Remark**
By the preceding remark we can identify the sections of $\mathcal{O}(\Omega)$ with analytic functions. On the other hand, the construction of $\mathcal{O}(\Omega)$ was based on analytic functions. There is a general result regarding the correspondence of a sheaf and its spaces of sections, see [Ten75] Chapter 2 (especially 5.7 Terminology) for a detailed explanation. In particular, the topology chosen in Construction B.1.5 occurs in a more general context.

In particular, the following sheaves can be constructed analogously to Construction B.1.5.

**B.1.12 Construction** (*The Sheaf of Germs of Analytic Sections of a Bundle*)
Let $\langle \mathcal{E} \overset{p}{\succ} \Omega \rangle$ be a bundle. If we substitute in Construction B.1.5 $A(O)$ with $\Gamma(O, \langle \mathcal{E} \overset{p}{\succ} \Omega \rangle)$ for each $\emptyset \neq O \overset{\circ}{\subset} \Omega$, we obtain a sheaf called the *sheaf of germs of sections (of the bundle $\langle \mathcal{E} \overset{p}{\succ} \Omega \rangle$)* and it is denoted by[2] $\mathcal{O}^{\langle \mathcal{E} \overset{p}{\succ} \Omega \rangle}(\Omega)$. In particular, for a fixed $z \in \Omega$ two sections of a bundle $s_i \in \Gamma(O_i, \langle \mathcal{E} \overset{p}{\succ} \Omega \rangle)$ for each $i = 1, 2$ are equivalent if there is a neighborhood $O$ of $z$, such that $O \subset O_1 \cap O_2$ and $s_1 = s_2$ on $O$. Thus $a \in \mathcal{O}^{\langle \mathcal{E} \overset{p}{\succ} \Omega \rangle}(\Omega)$ iff there is $a \in O \overset{\circ}{\subset} \Omega$ and $s \in \Gamma(O, \langle \mathcal{E} \overset{p}{\succ} \Omega \rangle)$ such that $a$ is the residue class of $s$. Again, the topology of $\mathcal{O}^{\langle \mathcal{E} \overset{p}{\succ} \Omega \rangle}(\Omega)$ can be described by the open sets $\{s \lhd O\} := \{\, \gamma_z(s) : z \in O \,\}$ for all $O \overset{\circ}{\subset} \Omega$ and $s \in \Gamma(O, \langle \mathcal{E} \overset{p}{\succ} \Omega \rangle)$.

**B.1.13 Construction**
Let $X$ be a Banach space. If we substitute in Construction B.1.5 $A(O)$ with

[2] Again, by abuse of notations the symbol is used both for the sheaf and the underlying topological space.

$A(O,X)$ for each $\emptyset \neq O \mathring{\subset} \Omega$, we obtain a sheaf $\mathcal{O}^X(\Omega)$. Clearly $\mathcal{O}^{\mathbb{C}}(\Omega) = \mathcal{O}(\Omega)$ and once we have introduced the sum of sheaves, cf. Definition B.1.22, we obtain that $\mathcal{O}^{\mathbb{C}^n}(\Omega)$ coincides with $\mathcal{O}^{\mathbb{C}}(\Omega))^n$.

**B.1.14 Remark**
Analogously to Remark B.1.10 we can identify

$$\Gamma(O, \langle \mathcal{E} \overset{p}{\succ} \Omega \rangle) \text{ with } \Gamma(O, \mathcal{O}^{\langle \mathcal{E} \overset{p}{\succ} \Omega \rangle}(\Omega))$$

and $A(O,X)$ with $\Gamma(O, \mathcal{O}^X(\Omega))$ where $\emptyset \neq O \mathring{\subset} \Omega$, $\langle \mathcal{E} \overset{p}{\succ} \Omega \rangle$ is a bundle and $X$ a Banach space.

**B.1.15 Definition**
For each $O \mathring{\subset} \Omega$ $\Gamma(O, \mathcal{O}(\Omega))$ is endowed with the Fréchet space structure of $A(O)$ (cf. Fact 1.5.6) via the bijection given by Remark B.1.10.

**B.1.16 Fact and Definition**
Let $\mathcal{F}$ be a sheaf over $\Omega$. Assume that for each $\emptyset \neq O \mathring{\subset} \Omega$ $\Gamma(O, \mathcal{F})$ carries a Fréchet space structure such that

1. $\Gamma(O, \mathcal{F})$ is a Fréchet $\Gamma(O, \mathcal{O})$ module, i. e. the multiplication $\Gamma(O, \mathcal{O}) \times \Gamma(O, \mathcal{F}) \to \Gamma(O, \mathcal{F})$ is continuous and
2. for all $\emptyset \neq o \mathring{\subset} O$ the restriction map $\Gamma(o, \mathcal{F}) \ni s \mapsto s_{|o} \in \Gamma(o, \mathcal{F})$ is continuous.

Then $\mathcal{F}$ is called a Fréchet sheaf.

**B.1.17 Definition** (*Homomorphism of Sheaves*)
Let $\mathcal{F}_1$ and $\mathcal{F}_2$ be sheaves over $\Omega$ with projections $p_1$, $p_2$, resp.. A continuous map $H : \mathcal{F}_1 \to \mathcal{F}_2$ is called a *homomorphism* iff $p_2 \circ H = p_1$ and for each $z \in \Omega$ the restriction of $H$ to $(\mathcal{F}_1)_z$ is a $\mathcal{O}_z$-homomorphism between the modules $(\mathcal{F}_1)_z$ and $(\mathcal{F}_2)_z$.

**B.1.18 Definition and Fact** (*Isomorphism of Sheaves*)
A homomorphism $H : \mathcal{F}_1 \to \mathcal{F}_2$ that is bijective is called an *isomorphism*. In that case, $H^{-1} : \mathcal{F}_2 \to \mathcal{F}_1$ is an homomorphism of sheaves, cf. [Kul70] Satz I.3.1.

**B.1.19 Definition** (*Restriction of a Sheaf*)
Let $\mathcal{F}$ be a sheaf over $\Omega$ with projection $p$ and $\emptyset \neq O \mathring{\subset} \Omega$. Endow $p^{-1}(O)$ with the induced topology from the topological space $\mathcal{F}$. For each $z \in O$ endow $p^{-1}(\{z\})$ with the $\mathcal{O}_z$-module structure given by the identification with

$\mathcal{F}_z$. Then $\mathcal{F}_{|O} := p^{-1}(O)$ is a sheaf over $O$ with projection $p_{|\mathcal{F}_{|O}}$, called the *restriction of* $\mathcal{F}$.

**B.1.20 Definition** (*Subsheaf*)
Let $\mathcal{F}_1$ and $\mathcal{F}_2$ be sheaves over $\Omega$ with projections $p_1$, $p_2$, resp.. Furthermore, assume that the topological space $\mathcal{F}_1$ is an open subspace of $\mathcal{F}_2$, that $p_1$ is the restriction of $p_2$ to $\mathcal{F}_1$ and that for each $z \in \Omega$ $(\mathcal{F}_1)_z$ is a submodule of $(\mathcal{F}_2)_z$. Then $\mathcal{F}_1$ is called a *subsheaf* of $\mathcal{F}_2$.

In that case, for each $O \overset{\circ}{\subset} \Omega$ $\Gamma(O, \mathcal{F}_1)$ is a submodule of $\Gamma(O, \mathcal{F}_2)$.

**B.1.21 Definition** (*Quotient of Sheaves*)
Let $\mathcal{F}$ be a sheaf over $\Omega$ with projection $p$ and let $\mathcal{G}$ be subsheaf of $\mathcal{F}$. Let $\mathcal{H} := \dot{\bigcup}_{z\in\Omega} {}^{\mathcal{F}_z}\!/\!_{\mathcal{G}_z}$ be the disjoint union of the pointwise quotient modules and denote by $\tilde{p} : \mathcal{H} \to \Omega$ the canonical projection. For each $z \in \Omega$ we denote by $q_z : \mathcal{F} \to {}^{\mathcal{F}_z}\!/\!_{\mathcal{G}_z}$ the canonical projection and we define $q : \mathcal{F} \to \mathcal{H}$ by $\mathcal{F} \ni a \mapsto q_{p(a)}(a)$. Finally, we endow $\mathcal{H}$ with the quotient topology induced by $q$.
Then ${}^{\mathcal{F}}\!/\!_{\mathcal{G}} := \mathcal{H}$ is a sheaf over $\Omega$ with projection $\tilde{p}$, called the *quotient sheaf* of $\mathcal{F}$ and $\mathcal{G}$.
In that case, for each $O \overset{\circ}{\subset} \Omega$ $\Gamma(O, {}^{\mathcal{F}}\!/\!_{\mathcal{G}})$ coincides with the quotient module ${}^{\Gamma(O,\mathcal{F})}\!/\!_{\Gamma(O,\mathcal{G})}$, cf. [Kul70] Satz I.5.3. and the remark thereafter.

**B.1.22 Definition** (*Sum of Sheaves*)
Let $n \in \mathbb{N}$ and $\mathcal{F}_1, \ldots, \mathcal{F}_n$ be sheaves over $\Omega$. Let $\mathcal{G} := \dot{\bigcup}_{z\in\Omega} (\mathcal{F}_1)_z \oplus \cdots \oplus (\mathcal{F}_n)_z$ be the disjoint union of the pointwise direct sums and denote by $\tilde{p} : \mathcal{G} \to \Omega$ the canonical projection. We endow $\mathcal{G}$ with the induced topology of the topological product space $\mathcal{F}_1 \times \cdots \times \mathcal{F}_n$.
Then $\mathcal{F}_1 \oplus \cdots \oplus \mathcal{F}_n := \mathcal{G}$ is a sheaf over $\Omega$ with projection $\tilde{p}$, called the *(Whitney) sum* of the sheaves $\mathcal{F}_1, \ldots, \mathcal{F}_n$.
If $\mathcal{F} := \mathcal{F}_1 = \cdots = \mathcal{F}_n$ then we set $(\mathcal{F})^n := \mathcal{F}_1 \oplus \cdots \oplus \mathcal{F}_n$.

**B.1.23 Definition** (Ker $H$, Range)
Let $H : \mathcal{F} \to \mathcal{G}$ be an homomorphism. Then $\operatorname{Ker} H := \{ a \in \mathcal{F} : H(a) = 0_{p(a)} \}$ is a subsheaf of $\mathcal{F}$ (where $p$ denotes the projection of $\mathcal{F}$ and $0_z$ denotes the zero element of $\mathcal{F}_z$) and $\operatorname{Range} H := H(\mathcal{F})$ is a subsheaf of $\mathcal{G}$.

**B.1.24 Definition** (*Exact Sequences*)
Let $n \in \mathbb{N}$ and $\mathcal{F}_1, \dots, \mathcal{F}_n$ be sheaves over $\Omega$. We say there exists an exact sequence $\mathcal{F}_1 \longrightarrow \mathcal{F}_2 \longrightarrow \dots \longrightarrow \mathcal{F}_n$ if there exist homomorphisms $H_i : \mathcal{F}_i \to \mathcal{F}_{i+1}$ for each $i = 1, \dots, n-1$ such that $\operatorname{Range} H_i = \operatorname{Ker} H_{i+1}$ for all $i = 1, \dots, n-1$.

The sheaves that occur in this thesis share an additional property, called "coherence".

**B.1.25 Definition** (*BCAF sheaf*)
Let $\mathcal{F}$ be a Fréchet sheaf over $\Omega$. Assume that for each $z \in \Omega$ there exist an neighborhood $O \overset{\circ}{\subset} \Omega$ of $z$ and Banach spaces $X$ and $Y$ such that there exists an exact sequence of the form

$$\mathcal{O}^X(O) \longrightarrow \mathcal{O}^Y(O) \longrightarrow \mathcal{F}_{|O} \longrightarrow 0.$$

Then $\mathcal{F}$ is called a *Banach coherent analytic Fréchet sheaf*, or *BCAF sheaf* for short.

**B.1.26 Definition** (*Coherent sheaf*)
If in Definition B.1.25 both Banach spaces $X$ and $Y$ are of finite dimension for each $z \in \Omega$, then $\mathcal{F}$ is called a *coherent (analytic) sheaf.*

**B.1.27 Remark**
The above definitions are adjusted to the situation of analytic sheaves over $\Omega \overset{\circ}{\subset} \mathbb{C}$. In order to allow a comparison with more general definitions commonly found in the referenced literature, we outline the correlation with the general situation. We will omit details and refer to the cited references for precise definitions.

BCAF sheaves (locally) allow exact sequences

$$\mathcal{O}^{X_1}(O) \longrightarrow \mathcal{O}^{X_2}(O) \longrightarrow \dots \longrightarrow \mathcal{O}^{X_n}(O) \longrightarrow \mathcal{F}_{|O} \longrightarrow 0.$$

of arbitrary (finite) length, i.e. BCAF sheaves are BCAF sheaves in the sense of [Lei78] Definition 2.1. (This is a consequence of $\Omega \overset{\circ}{\subset} \mathbb{C}$, cf. [Lei78] Problem 2.4 and the reference therein.)

Coherence in the sense of Definition B.1.26 can be characterized by the requirement that both $\mathcal{F}$ and all corresponding sheaves of relations are finitely generated (see [Kul70] for the definitions of finitely generated sheaves and the sheaf of relations): This is a consequence of the requirement that the underlying rings of the module structure are the stalks of $\mathcal{O}(\Omega)$, cf. [Kul70] Satz 28.2 and the fact that $\mathcal{O}(\Omega)$ is coherent, cf. Oka's Coherence theorem [Kul70] Satz 28.7.

**B.1.28 Remark**
In the situation of Definition B.1.26 the requirement that $\mathcal{F}$ is a Fréchet sheaf can be dropped. Indeed, by [KK83] Theorem 55.5 for all $U \overset{\circ}{\subset} \Omega$ $\Gamma(U, \mathcal{F})$ can be endowed with a topology such that $\mathcal{F}$ is a Fréchet sheaf:

**B.1.29 Fact**
Let $\langle \mathcal{E} \overset{p}{\succ} X \rangle$ be a bundle. Then $\mathcal{O}^{\langle \mathcal{E} \overset{p}{\succ} \Omega \rangle}(\Omega)$ is a BCAF sheaf (Cf. [Bun68] Theorem 4.2).

**B.1.30 Fact** (*Induced Homomorphism of Banach Vector Bundles on Sheaves*)
Let $A : \langle \mathcal{E} \overset{p}{\succ} \Omega \rangle \to \langle \mathcal{F} \overset{q}{\succ} \Omega \rangle$ be a homomorphism of bundles. Let $a \in \mathcal{O}^{\langle \mathcal{E} \overset{p}{\succ} \Omega \rangle}(\Omega)$. Then by definition there is $O \subset \Omega$, $s \in \Gamma(O, \langle \mathcal{E} \overset{p}{\succ} \Omega \rangle)$ and $z \in O$ such that $a = \gamma_z(s)$. We set $A_{\mathcal{O}}(a) := \gamma_z(A_{\Gamma|O}(s))$.

Then $A_{\mathcal{O}} : \mathcal{O}^{\langle \mathcal{E} \overset{p}{\succ} \Omega \rangle}(\Omega) \to \mathcal{O}^{\langle \mathcal{F} \overset{p}{\succ} \Omega \rangle}(\Omega)$ is a well-defined homomorphism of sheaves, called *the induced (sheaf) homomorphism by A*.

Furthermore, if $A$ is Fredholm, then $\operatorname{Ker} A_{\mathcal{O}}$ and

$$\operatorname{Coker} A_{\mathcal{O}} := \mathcal{O}^{\langle \mathcal{F} \overset{p}{\succ} \Omega \rangle}(\Omega) \big/ \operatorname{Range} A_{\mathcal{O}}$$

are coherent. (See [Kuc93] Theorem 1.6.14.)

# Bibliography


[AB02] Wolfgang Arendt and Shangquan Bu, *The operator-valued Marcinkiewicz multiplier theorem and maximal regularity*, Math. Z. **240** (2002), no. 2, 311–343.

[All67] G. R. Allan, *On one-sided inverses in Banach algebras of holomorphic vector-valued functions*, J. Lond. Math. Soc. **42** (1967), 463–470.

[Ama95] Herbert Amann, *Linear and quasilinear parabolic problems*, Abstract Linear Theory, vol. 1, Birkhäuser Verlag, Basel, 1995.

[AR09] Wolfgang Arendt and Patrick J. Rabier, *Linear evolution operators on spaces of periodic functions*, Commun. Pure Appl. Anal. **8** (2009), no. 1, 5–36.

[Aub63] Jean-Pierre Aubin, *Un théorème de compacite*, C. R. Acad. Sci., Paris **256** (1963), 5042–5044 (French).

[Bos04] Siegfried Bosch, *Algebra*, fifth ed., Springer, Berlin / Heidelberg, 2004 (German).

[Bou87] Nicolas Bourbaki, *Topological vector spaces*, Elements of Mathematics, Springer, Berlin / Heidelberg / New York, 1987.

[Bun68] Lutz Bungart, *On analytic fiber bundles. I: Holomorphic fiber bundles with infinite dimensional fibers*, Topology **7** (1968), 55–68.

[Cha85] Soo Bong Chae, *Holomorphy and calculus in normed spaces*, Dekker, New York, 1985.

[Der72] V. I. Derguzov, *The mathematical investigation of periodic cylindrical waveguides I*, Vestn. Leningr. Univ. **27** (1972), no. 13, 32–40 (Russian).

[DS58] Nelson Dunford and Jacob T. Schwartz, *Linear operators part I: General theory*, Pure and applied mathematics, vol. 7, Interscience Publishers, New York, 1958.

[Dug70] James Dugundji, *Topology*, fifth ed., Allyn & Bacon, Boston, 1970.

[Edw65] Robert E. Edwards, *Functional analysis. Theory and applications*, Holt Rinehart and Winston, New York / Chicago / San Francisco / Toronto / London, 1965.

[FL94] Wolfgang Fischer and Ingo Lieb, *Funktionentheorie*, seventh ed., Vieweg, Braunschweig, 1994 (German).

[FW68] Klaus Floret and Joseph Wloka, *Einführung in die Theorie der lokalkonvexen Räume*, Lecture Notes in Mathematics, vol. 56, Springer, Berlin, 1968.

[GR65] Robert C. Gunning and Hugo Rossi, *Analytic functions of several complex variables*, Prentice-Hall, Englewood Cliffs (New Jersey), 1965.

[GR84] Hans Grauert and Reinhold Remmert, *Coherent analytic sheaves*, Springer, Berlin / Heidelberg, 1984.

[Gro53] Alexandre Grothendieck, *Sur certains espaces de fonctions holomorphes. I.*, J. Reine Angew. Math. **192** (1953), 36–64 (French).

[Heu92] Harro Heuser, *Funktionalanalysis*, third ed., Teubner, Stuttgart, 1992 (German).

[Hör61] Lars Hörmander, *Hypoelliptic differential operators.*, Ann. Inst. Fourier **11** (1961), 477–492.

[Hör67] ______, *An introduction to complex analysis in several variables*, repr. ed., van Nostrand, Princeton (New Jersey), 1967.

[HP57] Einar Hille and Ralph S. Phillips, *Functional analysis and semigroups*, rev. ed., American Mathematical Society, Providence (Rhode Island), 1957.

[Jac62] John David Jackson, *Classical electrodynamics*, Wiley, New York, 1962.

[Jän05] Klaus Jänich, *Topologie*, eighth ed., Springer, Berlin / Heidelberg, 2005 (German).

[Kat66] Tosio Kato, *Perturbation theory for linear operators*, Die Grundlehren der mathematischen Wissenschaften, vol. 132, Springer, Berlin, 1966.

[KK83] Ludger Kaup and Burchard Kaup, *Holomorphic functions of several variables. An introduction to the fundamental theory*, De Gruyter Studies in Mathematics, vol. 3, Walter de Gruyter, Berlin / New York, 1983.

[KKW06] Nigel Kalton, Peer Kunstmann, and Lutz Weis, *Perturbation and interpolation theorems for the $H^\infty$-calculus with applications to differential operators.*, Math. Ann. **336** (2006), no. 4, 747–801.

[Kuc93] Peter A. Kuchment, *Floquet theory for partial differential equations*, Operator Theory: Advances and Applications, vol. 60, Birkhäuser Verlag, Basel, 1993.

[Kul70] Rolf Kultze, *Garbentheorie*, Teubner, Stuttgart, 1970 (German).

[KW04] Peer C. Kunstmann and Lutz Weis, *Maximal $L_p$-regularity for parabolic equations, Fourier multiplier theorems and $H^\infty$-functional calculus*, Functional Analytic Methods for Evolution Equations, Lecture Notes in Mathematics, vol. 1855, Springer, Berlin / Heidelberg, 2004.

[Lei78] Jürgen Leiterer, *Banach coherent analytic Frechet sheaves*, Math. Nachr. **85** (1978), 91–109.

[Łoj91] Stanisław Łojasiewicz, *Introduction to complex analytic geometry*, Birkhäuser Verlag, Basel, 1991.

[Mit70] B. S. Mityagin, *The homotopy structure of the linear group of a Banach space*, Russ. Math. Surv. **25** (1970), no. 5, 59–103.

[Pal68] Victor P. Palamodov, *Differential operators in coherent analytic sheaves*, Math. USSR, Sb. **6** (1968), 365–391.

[Pal93] ———, *Harmonic synthesis of solutions of elliptic equation with periodic coefficients*, Ann. Inst. Fourier **43** (1993), no. 3, 751–768.

[Pru76] A. S. Prutkovskii, *A solution of the stationary Maxwell equations for cylindrical waveguides with an anisotropic filling*, J. Soviet Math. **6** (1976), no. 1, 58–63.

[Que01] Boto von Querenburg, *Mengentheoretische topologie*, third ed., Springer, Berlin / Heidelberg, 2001 (German).

[RS02] Reinhold Remmert and Georg Schumacher, *Funktionentheorie 1*, fifth ed., Springer, Berlin, 2002 (German).

[Sch73] Martin Schechter, *Principles of functional analysis*, second ed., Academic Press, New York, 1973.

[Sch80] Helmut H. Schaefer, *Topological vector spaces*, fourth ed., Springer, New York / Heidelberg / Berlin, 1980.

[Ste51] Norman Steenrod, *The topology of fibre bundles*, Princeton Mathematical Series, vol. 14, Princeton University Press, Princeton, 1951.

[Ten75] B. R. Tennison, *Sheaf theory*, Cambridge Univ. Pr., Cambridge, 1975.

[Trè67] François Trèves, *Topological vector spaces, distributions and kernels*, Academic Press, New York / London, 1967.

[Tri95] Hans Triebel, *Interpolation theory, function spaces, differential operators*, second ed., Barth, Heidelberg / Leipzig, 1995.

[Wer05] Dirk Werner, *Funktionalanalysis*, fifth ed., Springer, Berlin / Heidelberg, 2005 (German).

[Yos71] Atsushi Yoshikawa, *Fractional powers of operators, interpolation theory and imbedding theorems*, J. Fac. Sci., Univ. Tokyo, Sect. I A **18** (1971), 335–362.

[ZKKP75] M. G. Zaidenberg, S. G. Krein, P. A. Kuchment, and A. A. Pankov, *Banach bundles and linear operators*, Russ. Math. Surv. **30** (1975), no. 5, 115–175.

# List of Symbols

# Index